24 Fortschritte der chemischen Forschung
Topics in Current Chemistry

Electronic Structure of Organic Compounds

Springer-Verlag Berlin Heidelberg GmbH

ISBN 978-3-540-05540-2   ISBN 978-3-540-36870-0 (eBook)
DOI 10.1007/978-3-540-36870-0

Contents

Chemically Induced Dynamic Nuclear Polarization

Prof. Dr. Hanns Fischer

Physikalisch-Chemisches Institut der Universität Zürich

Contents

1. Introduction

Chemically *induced* *dynamic* *nuclear* polarization (*CIDNP*) describes the appearence of emission and enhanced absorption in high resolution *nuclear* *magnetic* *resonance* (*NMR*) spectra of *radical reaction products* taken during or shortly after the course of the reaction. Discovered in 1967 [1a,2a], the phenomenon has recently attracted considerable interest, since CIDNP effects provide useful information on radical reactions and radical properties:

1. Caused by magnetic interactions in *transient radical pairs* involved in the steps of radical formation and destruction, the CIDNP effects provide evidence for radical intermediates. They can often be observed when radicals are present too briefly or in too low concentrations to be detectable by other methods.

2. The pathways of product formation affect the nature of the CIDNP phenomena. Therefore they enable distinctions to be made between
products of geminate pair combinations,
encounters of independently formed radicals, and
radical transfer reactions.

3. The effects are also influenced by the modes of pair formation. Geminate radical pair products and transfer products of radicals escaping the pairs show different CIDNP patterns when pairs are formed from reactions of singlet or triplet state precursors, so that CIDNP can be used to determine the *spin multiplicities of pair precursors.*

4. CIDNP may appear during reactions where the products are chemically identical with reactants; in this case they demonstrate that reaction is occuring and indicate the nature of the radical intermediates.

5. From CIDNP patterns it is possible to derive magnetic properties of free radicals as magnitudes and signs of hyperfine coupling constants, g factors and nuclear relaxation times.

The fact that no new instruments are needed is a major enticement to try CIDNP experiments. Conventional high-resolution NMR instruments may be used. Special applications require only minor modifications.

This report outlines the development and present status of CIDNP. Sect. 2 gives a brief account of the experiments so far reported and the generalizations of reactions and effects. Theoretical formulations of the current radical pair explanation of CIDNP are presented in Sect. 3, in particular for the so-called high-field case. Sect. 4 applies this theory to the interpretation of CIDNP phenomena in several illustrative examples, and, finally, Sect. 5 is devoted to the discussion of a few pertinent questions. Most of the work described in the literature is mentioned. However, we are not aiming at a broad and complete survey of all aspects and prefer to emphazise the basic facts and applications.

2. Survey of Results and Techniques

One of the earliest examples of CIDNP is the proton resonance emission of benzene observed during thermolysis of dibenzoylperoxide (I) in cyclohexanone [1a] which proceeds predominantly via (1) (see Sect. 4.1).

$$\text{\raisebox{0pt}{⬡}-CO_2-O_2C-\raisebox{0pt}{⬡}} \rightarrow 2\,\text{\raisebox{0pt}{⬡}}-CO_2\cdot \xrightarrow{-2CO_2} 2\,\text{\raisebox{0pt}{⬡}}\cdot \xrightarrow[-2R\cdot]{RH} 2\,\text{\raisebox{0pt}{⬡}} \quad (1)$$

This emission develops after the insertion of a solution of I into the pre-heated probe of the spectrometer and may be followed for several minutes (Fig. 1). Later on it changes into a normal absorption peak as the reaction is completed. This time dependence indicates that the benzene is formed with excess populations of the energetically higher nuclear Zeeman levels, i.e. with a specific nuclear polarization, and that it relaxes to thermal equilibrium after formation.

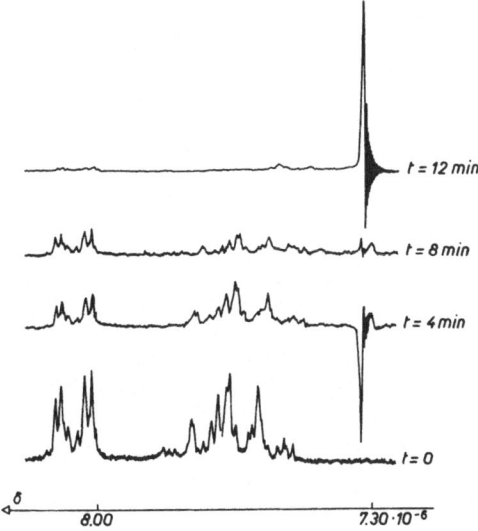

Fig. 1. CIDNP during thermolysis of dibenzoylperoxide in cyclohexanone (110 °C, 100 MHz)

Another example of early CIDNP phenomena is given in Fig. 2 (lower part). Emissions and enhanced absorptions appear in the vinyl proton resonances of 1-butene (A, B, C) and isobutylene (D) during the reaction of tert.-butyllithium with n-butylbromide [2a]

3

$$(CH_3)_3CLi + CH_3(CH_2)_3Br \longrightarrow LiBr + (CH_3)_2C=CH_2 + (CH_3)_3CH +$$
$$C_2H_5CH=CH_2 + (CH_3CH_2)_2 +$$

(2)

other products.

The upper part of Fig. 2 gives the spectrum after completion of the reaction.

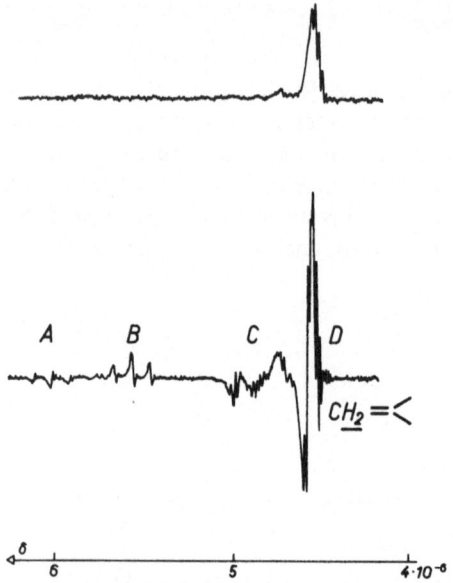

Fig. 2. CIDNP during the reaction of tert.-butyllithium with n-butylbromide

Superficially the effects of Figs. 1 and 2 may seem similar. There is, however, a major difference. The individual multiplets of Fig. 2 exhibit simultaneously emissions and enhanced absorptions of similar magnitudes, and the two effects nearly match within the multiplets. On the other hand, a net effect is found for the benzene line, which may be considered as being a degenerate multiplet. Thus, the two examples demonstrate two types of polarization which are useful to distinguish. *Multiplet effect* polarizations show either emission followed by enhanced absorption with increasing magnetic field (*EA*) (Fig. 2), or the reverse (*AE*). *Net effects* indicate emissions (*E*) (Fig. 1) or enhanced absorptions (*A*) for whole multiplets or single lines. As will be seen later, these two types represent the extremes of a broad range of observable CIDNP patterns.

First attempts to explain the new NMR phenomena [1b,2b] invoked electron-nuclear cross-relaxations in intermediate radicals and were based on a formalism similar to that of dynamic nuclear polarization or Overhauser effects [3]. Accordingly, the newly dicovered effect was called CIDNP. Though the initial ideas were capable of explaining the benzene emission [1b,d], it was soon realized that they were unable to account for most of the other experimental results, and today they are of historical interest only.

Nevertheless, the early papers stimulated subsequent work, in particular because of the basic assumption [1a,2a] that CIDNP effects relate to products of radical reactions only, and that they give evidence for radical intermediates. Up to now CIDNP effects have been reported for the following reaction types:

1. Thermal decompositions of acyl peroxides [1a,c,g;2e;4a,c,d;5a-h,8g], peresters [5a,b,d] azo compounds [1c,d;5g;6c,e,f] and N-nitrosohydroxylamine [9],

2. Photolysis of acyl peroxides [1e,f;4b;7],

3. Reactions of metal organic compounds with alkyl halides, such as alkyllithium-alkylhalide reactions [2a,c,d,f,h,i;8a-d;10], and reactions of Grignard compounds [2g] and sodium-naphthalenide [11] with alkylhalides,

4. Reductions of diazonium salts [12,13],

5. Molecular rearrangements involving 1, 2-, 1, 3- and 1,4-substituent shifts [8e,f;14a-e;15a-c;16a-c;17a;18]

6. Insertion reactions of triplet [6a,c,e] and singlet [19] carbenes,

7. Photoreductions of aromatic ketones and aldehydes [6b,d,g;20],

8. Norrish type I photocleavage of aliphatic ketones [21,22,23], and a few other systems [17b-d;24;25].

Of these reactions, groups 1, 2, 6, 7 and 8 are established radical reactions. As regards the others, the CIDNP evidence for radical intermediates was later supported by kinetic and ESR studies [13;14d,e;26]. Thus, the basic assumption that CIDNP is evidence for radicals is well confirmed and probably beyond question today.

The CIDNP experiments carried out on the thermal reactions 1, 3, 4 and 5 simply involved heating the reactants within the probes of NMR spectrometers or rapid mixing of reactants before insertion of the sample tubes. During the studies of photochemical reactions, simple modifications of spectrometer probes were introduced to permit irradiation of the reactants within the probes. These simple modifications have been described for 60 Mc/sec [6a] and for 100 Mc/sec [1f] - instruments. Several authors have used flow systems [1e] or specially designed all-quartz probes [27] instead. In most of the systems no special techniques are required for fast recording [1a]. The overwhelming majority of published examples deal with proton polarizations, an it is only recently that CIDNP effects have been reported for $^{13}C-$ [5f,g], $-^{15}N-$ [5g], $^{19}F-$ [1c,10] and $^{31}P-$ [25] resonances.

As far as the chemistry of the reactions leading to CIDNP is concerned, it appears that all the various reactions can be reconciled with Scheme I. Intermediate pairs of radicals in close proximity are formed by either unimolecular decompositions or bimolecular reactions of precursor molecules, or by random encounters of freely diffusing radicals. These pairs then either collapse to give com-

5

H. Fischer

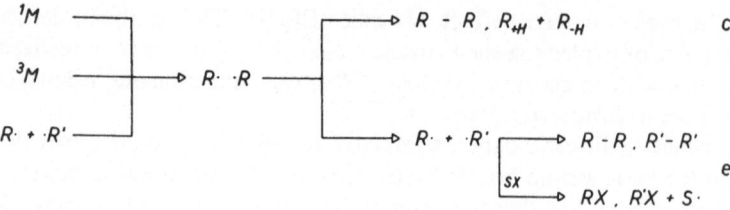

Scheme I

bination or disproportionation products, or separate into free radicals. Subsequently, the free radicals escaping from the pairs may be scavenged by suitable agents (*SX*) or may lead to other radical-radical reaction products after entering other pairs. It will be obvious later on why we have distinguished three types of pair formation in Scheme I, namely pair formations from reactions of precursors from electronic singlet states (*S*), from electronic triplet states (*T*), and by random radical encounters (*F*). It may also be noted that for *S* and *T* precursors the products formed by pair collapse (*c*), i.e. by encounters of the geminate radicals, may often properly be called "cage" products. On the other hand, we will here denote the products of radicals escaping the pairs as "escape" products (*e*).

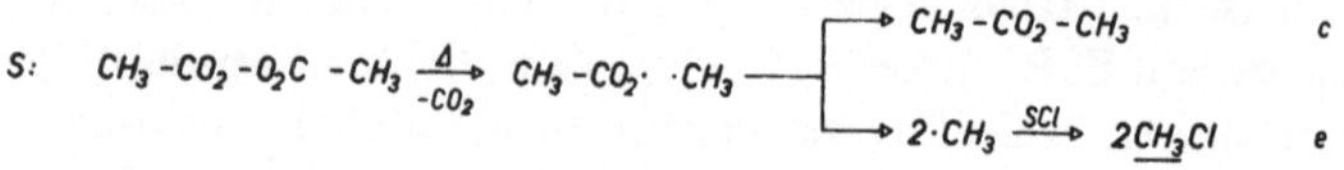

S: $CH_3 - CO_2 - O_2C - CH_3 \xrightarrow[-CO_2]{\Delta} CH_3 - CO_2 \cdot \ \cdot CH_3$

$\longrightarrow CH_3 - CO_2 - CH_3$ *c*

$\longrightarrow 2 \cdot CH_3 \xrightarrow{SCl} 2CH_3Cl$ *e*

Scheme II [4a]

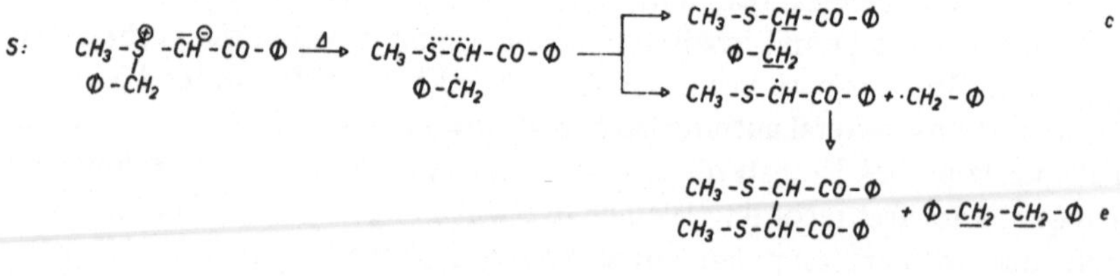

S: $CH_3 - \overset{O}{\underset{\Phi - CH_2}{\overset{\|}{S}}} - \overset{\ominus}{CH} - CO - \Phi \xrightarrow{\Delta} CH_3 - \overset{\cdots}{S} - \overset{\cdot}{CH} - CO - \Phi$
$\Phi - CH_2$

$\longrightarrow CH_3 - S - \underset{|}{CH} - CO - \Phi$ *c*
$\qquad \Phi - \overset{\cdot}{C}H_2$

$\longrightarrow CH_3 - S - \overset{\cdot}{CH} - CO - \Phi + \cdot CH_2 - \Phi$

$\downarrow$

$CH_3 - S - CH - CO - \Phi$
$CH_3 - S - \underset{|}{CH} - CO - \Phi \quad + \Phi - \underline{CH_2} - \underline{CH_2} - \Phi \ e$

Scheme III [14a]

T: $\Phi - CH_2 - CO - CH_2 - \Phi \xrightarrow{h\nu} \Phi - CH_2 - \overset{O}{\overset{\|}{C}} \cdot \ \cdot CH_2 - \Phi$

$\longrightarrow \Phi - CH_2 - CO - \underline{CH_2} - \Phi$ *c*

$\xrightarrow{-CO} 2 \ \Phi - \overset{\cdot}{C}H_2 \longrightarrow \Phi - \underline{CH_2} - \underline{CH_2} - \Phi \ e$

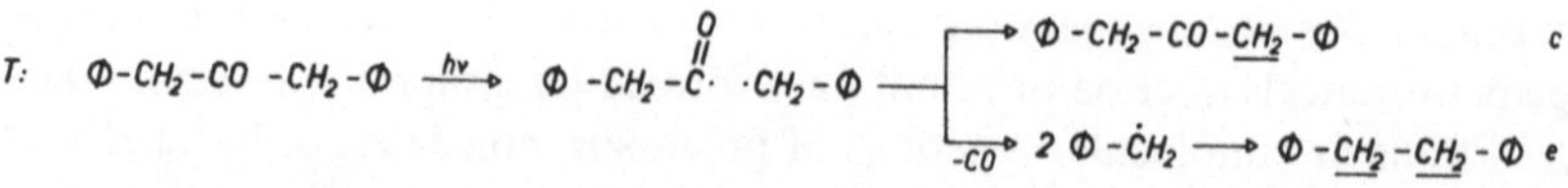

Scheme IV [23]

6

In general, CIDNP effects have been observed in NMR transitions of the products of both types (c) and (e) simultaneously. They are found, for instance, for the underlined nuclei in Schemes II to V, which are examples of Scheme I.

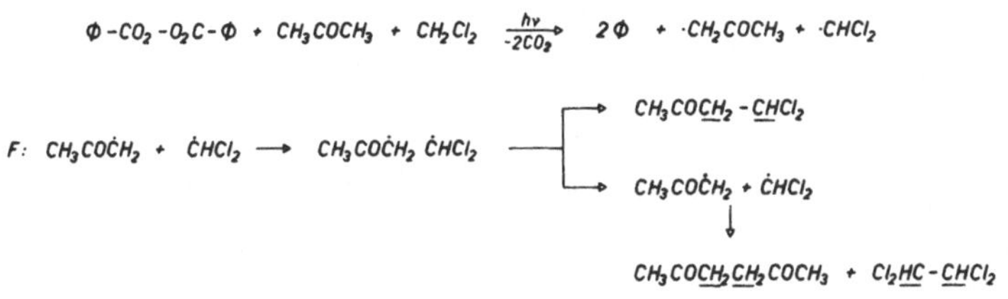

$$\phi-CO_2-O_2C-\phi \ + \ CH_3COCH_3 \ + \ CH_2Cl_2 \ \xrightarrow[-2CO_2]{h\nu} \ 2\phi \ + \ \cdot CH_2COCH_3 \ + \ \cdot CHCl_2$$

$$F: \ CH_3CO\dot{C}H_2 \ + \ \dot{C}HCl_2 \ \longrightarrow \ CH_3CO\dot{C}H_2 \ \dot{C}HCl_2 \ \left\{ \begin{array}{l} \longrightarrow \ CH_3CO\underline{CH}_2-\underline{CHCl}_2 \\ \\ \longrightarrow \ CH_3CO\dot{C}H_2 \ + \ \dot{C}HCl_2 \end{array} \right.$$

$$\downarrow$$

$$CH_3CO\underline{CH}_2\underline{CH}_2COCH_3 \ + \ Cl_2\underline{HC}-\underline{CHCl}_2$$

Scheme V [11]

If many different products are formed, the CIDNP patterns may be quite complex. They may also be of the net or multiplet effect types, or they may represent a superposition of both effects. However, two simplifying rules for CIDNP phenomena have been found experimentally. They serve as starting points for the interpretation and hint at the origin of the effects:

A. The polarizations of "cage" products (c) are related to those of "escape" products (e) for a given system. If a nucleus belongs to group R (or R') in the "cage" product and shows emission E or enhanced absorption A for a specific transition, then the corresponding transition of the same nucleus in the "escape" product shows the opposite polarization. Thus, in Scheme II, the CH_3 protons show E for $CH_3CO_2\underline{CH}_3$ and A for $\underline{CH}_3Cl$, in Scheme III the benzylic protons give E for the rearrangement product, A for dibenzyl, and in Scheme V the $CHCl_2$ proton gives E for the unsymmetric coupling product and A for tetrachloroethane, whereas the $\underline{CH}_2-COCH_3$ protons show A for the unsymmetric coupling and E for the symmetric coupling products.

B. The polarizations of the reaction products are related to the modes of pair formation. They are opposite for the cases of pair formation from S and T precursors and are equal in signs for T and F precursors. An example of this relationship will be given in Sect. 4.1.

Both relationships have been established experimentally [4a,6c]. They indicate that the common intermediate of Scheme I, the radical pair, is the origin of the effects and rule out the possibility that interactions within the freely diffusing uncorrelated radicals play major roles.

3. Theoretical Considerations

3.1 Enhancement Factors

Before the current CIDNP is stated, it is necessary to define the quantities which are generally employed in quantative descriptions of signal enhancements.

7

The intensity of an non-degenerate NMR transition between a lower nuclear Zeeman level K and a higher level L of a molecule is proportional to the difference in level populations [28]

$$I_{KL} \sim n_K - n_L \tag{3}$$

and in thermal equilibrium

$$I_{KL}^0 \sim n_K^0 - n_L^0 = \frac{n}{Z} \frac{g_n \beta_n H_0}{kT} \tag{4}$$

where n is the number of molecules present, and $Z = \prod_i (2I_i + 1)$ is the total number of levels.

NMR emission corresponds to $I_{KL} < 0$ and enhanced absorption to $I_{KL} > I_{KL}^0$. It is convenient to express the observed deviations from thermal equilibrium by the enhancement factor

$$V_{KL} = \frac{I_{KL} - I_{KL}^0}{I_{KL}^0} \tag{5}$$

By measuring the intensities I_{KL} in CIDNP patterns and the values I_{KL}^0 corresponding to thermal equilibrium of the same system, V_{KL} can be determined experimentally. The usual procedure for the determination of I_{KL}^0 involves rapid quenching of the reaction and allowance for thermal equilibration.

In general, it will be desired to relate the enhancement factors V_{KL} to the rates of product formation in levels K and L. This requires a rather involved procedure since the level populations n_K and n_L change with time, not only by the desired rates r_K and r_L, but also by relaxation, and in principle the set of coupled differential equations

$$\dot{n}_K = r_K - \sum_L^Z w_{KL} \{(n_K - n_L) - (n_K^0 - n_L^0)\} \tag{6}$$

has to be solved for all K. In (6) w_{KL} is the relaxation probability between levels K and L. To avoid the necessary integrations of (6), and since the parameters w_{KL} are often not known, many authors have applied approximations: the system is assumed to be describable by a quasi-steady state ($\dot{n}_K = 0$) and by one relaxation time $T_1 = \dfrac{1}{Z w_{KL}}$ for all K and L. With these assumptions, V_{KL} is approximated by

$$V_{KL} = T_1 \frac{r_K - r_L}{n} \frac{ZkT}{g_n \beta_n H_0} \tag{7}$$

Replacing r_K by rp_K, where r is a chemical rate of product formation and p_K is the probability for population of level K,

$$V_{KL} = T_1 \cdot \frac{r}{n} \cdot \underline{V_{KL}} = T_1 \cdot \frac{r}{n} \cdot (p_K - p_L) \frac{ZkT}{g_n \beta_n H_0} \qquad (8)$$

The theoretical enhancement factor *per product molecule formed* $\underline{V_{KL}}$ was first introduced by Closs [6] and is commonly used now to describe the signal enhancements in quantitative terms.

One should keep in mind, however, that (7) and (8) are only approximations for the relation between experimentally determined enhancement factors and probabilities of level populations. The use of these formulae may render comparisons of calculated and observed CIDNP patterns inconclusive unless the relaxation of a multi-level spin system can be described, at least approximately, by a single relaxation time [28]. To some extent this difficulty may be removed by attributing different relaxation times $T_{1,KL}$ to different transitions, but even then intramolecular Overhauser effects may still cause deviations of observed from calculated intensity distributions, especially within multiplets [1f]. Nevertheless, for many applications the approximate formulae (7) to (8) seem adequate.

3.2 The Radical Pair Mechanism

On the basis of the idea that CIDNP effects are caused by interactions within radical pairs, quantitative formalisms capable of explaining many features of CIDNP were given by Closs [29] and by Kaptein and Oosterhoff [30]. The theory was later refined and modified by various authors, but the new developments did not change the basic concepts.

Consider two free radicals formed in close proximity to one another by any one of the pathways S, T and F introduced in Sect. 2, and consider the probability of finding this pair in a state of pure electron multiplicity at the instant of formation. It is quite natural to suppose that the initial state is singlet if the pair is formed from a singlet precursor S, and that one of the three triplet states is occupied if the precursor is triplet (T). For the case of pair formation by random encounters of radicals (F), the singlet and the three triplet states will be occupied with equal probilities. After pair formation, the interradical distance will change because of diffusive displacements. The radicals will diffuse apart but there will be a finite probability that the radicals will reencounter after some time. During reencounters, some of the pairs will react to pair products by combination or disproportionation, others will separate again. Finally, the radicals may react with suitable agents or with other radicals during the times of random walk. These processes of pair reencounter reactions and of competitive scavenging give the products denoted c and e in Scheme I.

Two further assumptions now give the CIDNP effects if combined with the dynamic pair behavior described above. They are:

1. Pairs may undergo transitions between singlet and triplet states *with nuclear spin-dependent probabilities* in the time between successive encounters.

2. At radical-radical encounters, reaction to products c occurs for singlet state pairs only. Triplet state pairs do not react and separate instead.

It is easy to see how CIDNP arises from these assumptions. Consider a pair formed in a T state. For *specific* nuclear spin configurations, this pair will undergo rapid intersystem crossing to become an S pair during the random walk following the formation. For other nuclear spin configurations, it remains T for longer times. At reencounters, reaction thus occurs predominantly for specific nuclear spin configurations, and the pair reaction products (c) are formed with these nuclear spin configurations preferentially populated. On the other hand, the radicals of the pairs remaining in triplet states will have predominantly nuclear spin configurations which do not allow rapid intersystem crossing. Since these pairs have a higher probability of leading to products e during the random walk, the reason for rule A of Sect. 2 becomes obvious. Rule B also easily follows from a comparison of S and T precursor pair behaviour.

The forces which drive the intersystem crossing are the nuclear spin-dependent hyperfine interactions in the radicals and the electron Zeeman interactions. This becomes evident from the following: after pair formation in the magnetic field of a NMR spectrometer, say, the two unpaired electron spins precess about the magnetic field axis starting from defined initial phase angles. These initial phase angles are different for the four possible initial electronic states T_+, T_0, T_-, and S. They are given in Fig. 3 for three of these states. During the lifetime of the pairs, the precession keeps phase provided the precession frequencies of the two electron spins are exactly equal. Now the precession frequency, or Larmor frequency, of a radical in a magnetic field is given for high fields by the g factor and the hyperfine interactions with the nuclei [35] and is to first order (angular frequency units)

$$\omega_L = g\beta H_0 + \sum_\lambda a_\lambda m_{I\lambda} \tag{9}$$

where β is the Bohr magneton and a_λ are the hyperfine coupling constants (angular frequency units) of the electron-nuclear interactions. If ω_L is different for the two radicals, the precessions lose their initial phase relations and the S pairs will assume some T character, for instance. From (9) it is easily seen that the magnitude of this intersystem crossing will depend on the difference $g-g'$ of the g factors, on the hyperfine interactions a_λ and a_λ', and on the nuclear spin configurations $m_{I\lambda}$ and $m_{I\lambda}'$ of the two radicals $R\cdot$ and $R\cdot'$.

Fig. 4 shows another approach to the same problem. Here the energies of the four electronic pair states are plotted versus the radical pair distance for a certain hypothetical nuclear spin configuration and for high magnetic field. The

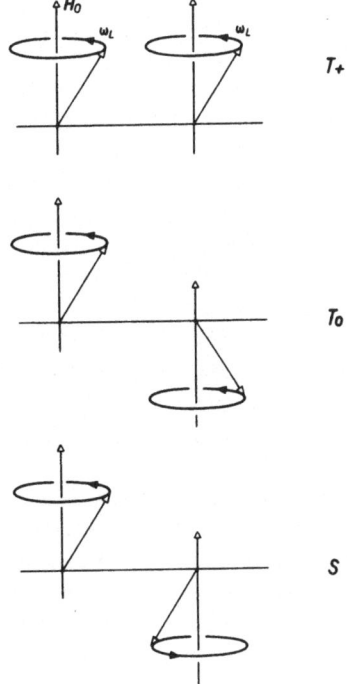

Fig. 3. Precession of electron spins

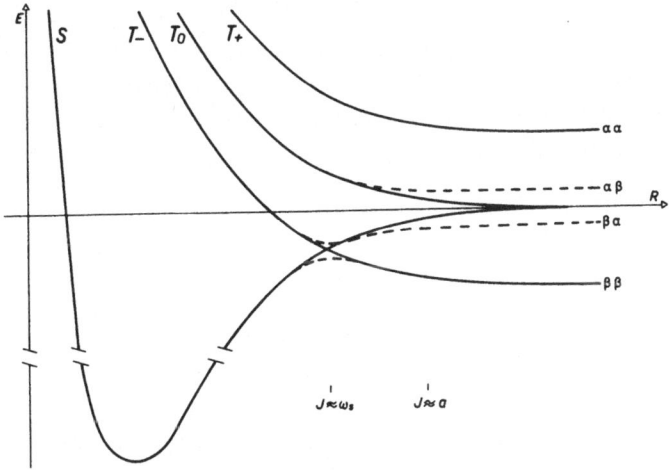

Fig. 4. Schematic representation of the electric energies of a radical pair with a hypothetical nuclear spin configuration vs. inter-radical distance

11

exchange interaction between the unpaired electrons is also taken into account. For small distances the singlet is lowest and the three triplet states are not degenerate because of the different Zeeman energies. For large distances, the pure spin states (full lines) cross or approach and become degenerate. Now g factor differences and hyperfine interactions tend to lift these degeneracies (broken lines) and to mix the states. Mixing occurs as the system travels through these mixing regions by diffusive displacements. Pairs starting from S states at low distances have adopted some T character when returning to encounter, and the degree of T character is again determined by the parameters $g-g'$, a_λ and $m_{I\lambda}$.

Fig. 4 also shows that the regions of mixing are larger for S and T_0 states than for S and T_- states for high magnetic fields.

Therefore, we expect the nuclear spin dependent $S-T_0$ transitions to predominate in the high-field case, whereas in low fields transitions between all states may become of equal importance. Further, we may also anticipate that the time dependence of the interradical distance will influence the transition rates.

The following theoretical considerations apply especially for the high-field case of generation and observation of CIDNP effects with fields larger than about 1000 gauss. Low-field polarizations will be mentioned in Sect. 5.

There are various quantitative descriptions of the pair mechanism [29-34] which are based on different treatments of the time dependences. For high-field polarizations, at least, they all lead to similar agreements of observed with calculated CIDNP patterns.

A common starting point is the high-field spin Hamiltonian of a radical pair $R \cdot \cdot R'$ (angular frequency units)

$$\mathcal{H} = \frac{J}{2}(1 + 4\vec{S}\vec{S'}) + g\beta H_0 \vec{S}_z + g'\beta H_0 \vec{S}_z' + \sum_\lambda a_\lambda \vec{S}_z \vec{I}_{z\lambda} + \sum_{\chi'} a_\lambda' \vec{S}_z' \vec{I}_{z\lambda'} \qquad (10)$$

which contains the electron exchange interaction (J), the electron Zeeman interactions and the isotropic electron-nuclear hyperfine interactions. Electron-electron and electron-nuclear dipole-dipole terms are usually omitted. All treatments also use the high-field representation of the electron-nuclear pair states

$$S\chi_k = \frac{1}{\sqrt{2}}(\alpha\beta' - \beta\alpha')\chi_k$$

$$T_+\chi_k = \alpha\alpha'\chi_k$$

$$T_0\chi_k = \frac{1}{\sqrt{2}}(\alpha\beta' + \beta\alpha')\chi_k \qquad (11)$$

$$T_-\chi_k = \beta\beta'\chi_k$$

The χ_k are the simple product functions of the nuclear spins in the pairs

$$\chi_k = \prod_\lambda |m_{I\lambda} > \prod_{\chi'} |m_{I\lambda'} > \qquad (12)$$

$\mathcal{H}$ has matrix elements beween S and T_0 states of the same nuclear spin configurations k only, and these are given by

$$\Omega_k = \frac{1}{2} \{(g-g') \beta H_0 + \sum_\lambda a_\lambda m_{I\lambda} - \sum_{\chi} a_{\chi'} m_{I\chi'}\} \tag{13}$$

It should be noted that the nuclear spin states of the pairs (12) may be identified with those of the pair products c which were previously denoted K and L in cases of simple first-order NMR spectra of the products. In other cases, the coefficients of the expansion of the spin functions χ_K of the products in terms of the χ_k

$$\chi_K = \sum_k c_{Kk} \chi_k \tag{14}$$

are often available.

In the original approach by Closs and Kaptein-Oosterhoff [29,30] the time-space behaviour of a pair is approximated by the following model: a pair formed at $t = 0$ in one of the states (11) remains for some time t in regions of inter-radical distances where the exchange integral J is of the order of magnitude of Ω_k and is represented by an average fixed value $\bar{J}$. At time t the pair either reacts to product c or separates to distances of no return. Product c is formed only if the pair state is singlet at time t. An exponential distribution of life times $f(t)$ was proposed

$$f(t) = \frac{1}{\tau} e^{-t/\tau} \tag{15}$$

τ being an average lifetime. Expressing the probability of finding the pair in a singlet state with nuclear spin configuration k as $p_{S,k}(t)$, the probability of forming the product c in the same nuclear spin configuration k becomes

$$p_k (c) \sim \int_0^\infty p_{S,k}(t) \cdot f(t) \, dt \tag{16}$$

For first-order NMR spectra of product c, this is the quantity p_K needed for estimates of the enhancement factors (8). Otherwise p_K follows from (14) and (16) as [1c]

$$p_K = \sum_k |c_{Kk}|^2 \cdot p_k \tag{17}$$

Now $p_{S,k}(t)$ is calculated from the time-dependent Schrödinger equation

$$\mathcal{H} \psi_k = i \frac{\partial \psi_k}{\partial t} \tag{18}$$

with the expansion

$$\psi_k (t) = (c_{S,k} (t)S + c_{T_0,k}(t) T_0) \chi_k \tag{19}$$

as

$$p_{S,k} = |c_{S,k}|^2 \tag{20}$$

Solution of (18) gives

$$c_{S,k} (t) = c_{S,k} (0) \cos \omega_k t - \frac{i}{\omega_k} (\bar{J} \cdot c_{S,k} (0) + \Omega_k \cdot c_{T_0,k}(0)) \sin \omega_k t \tag{21}$$

13

where $c_{S,k}(0)$ and $c_{T_{0,k}}(0)$ are the initial coefficients determined by the precursor states and where

$$\omega_k = (\bar{J}^2 + \Omega_k^2)^{1/2} \tag{21}$$

Integration of (16) then gives the pertinent quantities $p_k(c)$

$$p_k^s(c) \sim \frac{2\Omega_k^2 \tau^2}{1+4\omega_k^2 \tau^2} \tag{22}$$

$$p_k^T(c) \sim \frac{2\Omega_k^2 \tau^2}{1+4\omega_k^2 \tau^2} \tag{23}$$

for the cases of singlet and triplet precursors.

For free radical precursors, the value of $p_k^F(c)$ should be that of triplet precursors, only of smaller magnitude [6e)]

$$p_k^F(c) \sim p_k^T(c) \tag{24}$$

because encounters of radicals with uncorrelated electron spins will initially populate S and T states alike and some of the singlet pairs will react during encounters. Thus, the remainder behave as if they possessed predominantly T precursors.

The probabilities of "escape" product formation are related to those of pair product formation by

$$r_k(e) = r(e) \cdot p_k(e) = \kappa \cdot (r_k - r(c) \cdot p_k(c)) \tag{25}$$

where r_k is the rate of pair formation in level k, and $r(e)$, $r(c)$ are the total rates of c and e-product formation. The factor κ includes terms of reaction and relaxation of the separated radicals and will be treated later on (Sect. 4.3).

From (8), (17), (22), (23), (24), and (25) two relations arise between the enhancement factors of a given NMR transition of a product formed via the various pathways

$$V_{KL}^S(c) \sim - V_{KL}^T \sim - V_{KL}^F(c) \tag{26}$$

$$V_{KL}(e) \sim - V_{KL}(c) \tag{27}$$

They are identical with the rules stated in Sect. 2.

Recently Kaptein [34)] and Adrian [32)] have modified the original model by introducing a more adequate function $f(t)$. Noyes [36)] has treated the diffusive behaviour of radical pairs and has calculated the probability of the first reencounter beween t and $t + dt$ for a pair separating at $t = 0$ from an encounter to be

$$f(t) dt \approx c \cdot t^{-3/2} dt \tag{28}$$

for long times. Using (28) in (16), Kaptein [34] obtains for the quantities $p_k(c)$ as approximations

$$p_k^S(c) = c_1 - c_2 \frac{\Omega_k^2}{\omega_k^{3/2}} \tag{29}$$

$$p_k^T(c) = c_3 \frac{\Omega_k^2}{\omega_k^{3/2}} \tag{30}$$

$$p_k^F(c) = c_4 + c_5 \frac{\Omega_k^2}{\omega_k^{3/2}} \tag{31}$$

where the constants c_1 to c_5 contain reaction and random walk parameters. An effective exchange integral $\bar{J}$ is retained in ω_k (20). Adrian [32] has argued that $\bar{J}$ should be zero since most of the polarization is built up in pairs which separate far apart before reencounters, and hence his treatment leads to formulas nearly identical with (29) to (31) if ω_k is replaced by Ω_k.

The polarizations of the products e of reactions competing with pair product formation have also been treated by Kaptein [34] and again relation (25) was found to hold.

The approaches of Closs, Kaptein-Oosterhoff, and Adrian are based on quantitative treatments of the microscopic behaviour of radical pairs. Very similar results can also be obtained from a simple kinetic model which involves formal rate constants [31b] for the processes of pair reaction from singlet states k_c, pair "escape" k_d, and singlet-triplet transitions $k_{isc,k}$. Replacement of the actual pair behaviour by a simple kinetic scheme may be an oversimplification, though it seems justified by the results of Szwarc and co-workers [37] who showed that pair reaction and escape may be described to a good approximation in terms of simple first-order processes.

With the approximation

$$k_{isc,k} \sim \Omega_k^2 \tag{32}$$

the kinetic formulation gives [31b]

$$p_k^S(c) \sim \frac{1}{1 + \dfrac{k_d}{k_d + k_c} \dfrac{\Omega_k^2}{K + \Omega_k^2}} \tag{33}$$

$$p_k^T(c) \sim \frac{\dfrac{\Omega_k^2}{K + \Omega_k^2}}{1 + \dfrac{k_d}{k_d + k_c} \dfrac{\Omega_k^2}{K + \Omega_k^2}} \tag{34}$$

$$p_k^F(c) \sim 1 + \frac{k_c}{k_d + k_c} \frac{\dfrac{\Omega_k^2}{K + \Omega_k^2}}{1 + \dfrac{k_d}{k_d + k_c} \dfrac{\Omega_k^2}{K + \Omega_k^2}} \tag{35}$$

and $p_k(e)$ (given by Eq. (25)), where K is a parameter which is of the order of magnitude of Ω_k^2. As is easily seen, this treatment gives the same relations for the enhancement factors as the previous ones.

In actual calculations of CIDNP spectra from the theories outline above, it is often required to compute relative line enhancements rather than absolute values. This is the reason why we have omitted proportionality constants in listing the expressions (22) to (24), (29) to (31), and (33) to (35) for the probabilities p_k. Another reason is that none of the treatments gives precise theoretical values for the proportionality constants.

The calculations start with evaluation of the matrix elements Ω_k from known g factors and hyperfine coupling constants of the radicals involved. In treatments involving more or less arbitrary parameters (τ, J, K), these are then chosen and the quantities p_k are calculated. If the products show first-order NMR spectra, the probabilities p_k are inserted directly into (8), and the desired NMR patterns result. Otherwise the transformation (17) has to be used to obtain the probabilities p_K from the p_k.

It should be mentioned that the parameters τ in Eq. (22), (23), J in (29)– (31) and K in (33) – (35) influence the calculated CIDNP patterns. In the literature, values of $\tau \approx 10^{-9}$ sec, $J \approx 10^8$ rad/sec [6,34] and $K \approx 10^{17}$ sec^{-2} [1f] were found appropriate.

Various comparisons of experimental and predicted CIDNP spectra have been reported, and an example is given in Fig. 5. Similar agreement is obtained for any of the above-mentioned theories of Closs-Kaptein-Oosterhoff [6] Adrian [32], Kaptein [34], and Fischer [31b]. This may be taken as support for the basic radical pair concept which is inherent in all the formulations.

As is seen from the equations for the probabilities p_k given above, the amplitude and sign of the enhancement factor of an NMR transition of a product depend on several parameters:

a) on the type of product ("cage" or "escape");
b) on the mode of pair formation (S, T, F precursor);
c) on the difference of g factors of the two radicals of the pair; and
d) on the hyperfine coupling constants.

Whereas accurate determinations of the latter parameters from CIDNP spectra will always require careful simulations, some of the others can also be derived from a simplified analysis which makes use of two simple rules for the signs of enhancement factors which are easily derived from equations (22), (23), (24)– (31), and (33)–(35) for the case of small g–g′ in Ω_k [31-34,38].

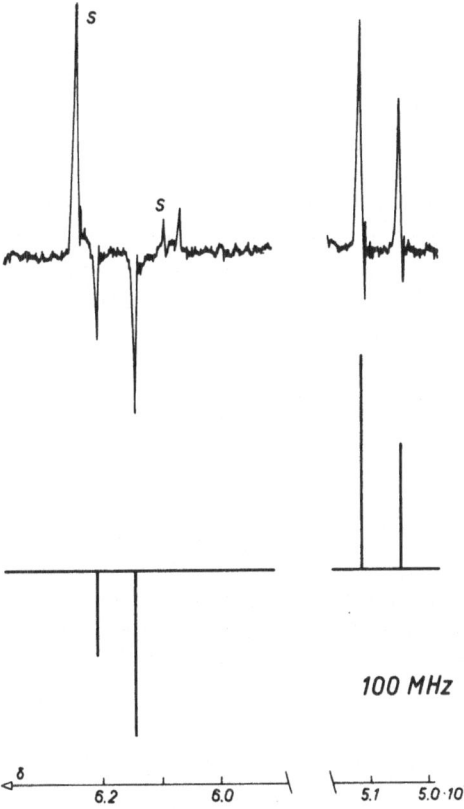

Fig. 5. Experimental and calculated CIDNP line intensities of $CHCl_2-CHCl-COOH$, observed during photolysis of dibenzoyl peroxide in a 3 : 1 mixture of CH_2Cl_2 and $CH_2ClCOOH$ [1f]. Lines denoted by S are solvent satellites. The line A at $\delta = 6.06$ ppm is due to $CHCl_2-CHCl_2$

Consider a product nucleus i (or a group of equivalent nuclei) which was part of the radical $R\cdot$ in the pair $R\cdot\ \cdot R'$, and which is coupled to nucleus j in the product with the nuclear spin coupling constant J_{ij}. The polarizations of the transitions of the NMR multiplet of this nucleus i may then be described as the superposition of a *net* effect and a *multiplet* effect, i.e. of two quantities Γ_{ni} and Γ_{mij} [39]

$$\Gamma_{ni} = \mu \cdot \epsilon(g-g')a_i \qquad (36)$$

$$\Gamma_{mij} = \mu \cdot \epsilon a_i a_j J_{ij} s_{ij} \qquad (37)$$

Positive Γ_{ni} corresponds to enhanced absorption, negative Γ_{ni} to emission of the multiplet. Positive Γ_{mij} corresponds to a multiplet effect of *EA* character, negative Γ_{mij} to *AE*. The parameters μ, ϵ and s_{ij} refer to the reactions involved and have the following signs:

17

μ is positive for T and F precursors, and negative for S precursors,

ϵ is positive for nucleus i residing in a "cage" combination or disproportionation product, and negative for "escape" products,

s_{ij} is positive if nuclei i and j belong to the same radical in R· ·R', and negative
if they belong to different radicals.

Illustrative examples of these convenient rules will be given in Sect. 4, and
a survey shows that most of the published CIDNP patterns agree with predictions based on (36) and (37). However, exceptions are also found [1f,40], and it
appears that (37) in particular may give erroneous results for high values of
$(g-g')\beta H_0$ in Ω_k (13). In such cases, a proper simulation of CIDNP patterns
from the full formulae is recommended.

4. Examples

4.1 CIDNP During Aroyl Peroxide Decompositions

As noted previously, the emission of benzene observed during thermolysis of
dibenzoylperoxide (I) was one of the first examples quoted of CIDNP effects.
It has been the subject of a recent quantitative study [1g]. The chemical reactions are given in Scheme VI. They follow from product distribution [1g,41].

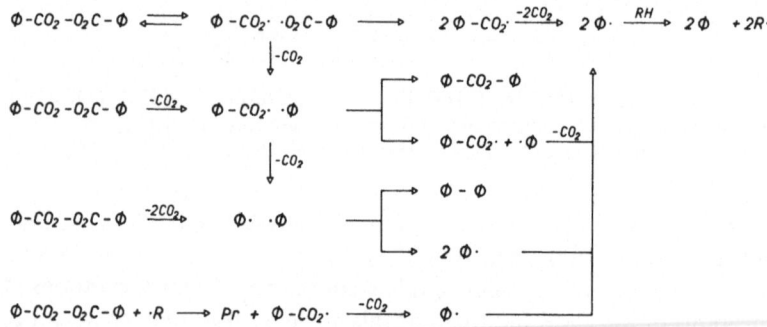

Scheme VI [1g]

Pr denotes the product of the induced decomposition, which may be $\emptyset$–CO_2–R,
where R is the radical derived from the solvent (RH). The yields of the "cage"
products phenylbenzoate and diphenyl amount to 4% and 0.1%, respectively.
In Scheme VI the benzene is seen to be formed by a transfer reaction of phenyl
radicals, and two types of phenyl radicals can be distinguished, namely those
arising by decarboxylation of benzoyloxy radicals and those escaping the primary pairs benzoyloxy/phenyl and phenyl/phenyl. Now, scavenging of the ben-

zoyloxy radicals with iodine and hydrolysis of the hypoiodide had no effect on the overall polarizations [1g]. Therefore the benzene emission results from phenyl radicals escaping from the pairs benzoyloxy/phenyl and phenyl/phenyl. Phenyl/ phenyl cannot be the polarizing system since the diphenyl was not found polarized, and this leaves interactions in benzoyloxy/phenyl pairs as the source of polarization. Eq. (36) then easily explains the benzene emission: the reaction is thermal and the pairs are formed from S precursors, i.e. μ is negative. Benzene as "escape" product corresponds to negative ϵ. From ESR, g factors of carbon radicals as phenyl are known to be smaller than g factors of oxy-radicals [35,42], thus $g-g'$ is negative. Finally, the hyperfine coupling constants of the protons of phenyl are positive [43], thus Eq. (36) reads for the benzene protons

$$\Gamma_{ni} = ---+ = - \qquad (38)$$

and emission is expected. On the other hand, enhanced absorption is predicted for the "cage" product phenylbenzoate ($\epsilon > 0$). In Fig. 1 a small transient enhanced absorption is seen on the high-field side of the benzene emission. This might be due to the phenylbenzoate which has complicated A_2B_2C NMR patterns. The assignment is likely, however, since the two enhanced absorption lines observed at $\delta = 7.00$ and $\delta = 7.09$ during thermolysis of di-(4-chlorobenzoyl)-peroxide in hexachloroacetone are unambiguously assigned to the corresponding 4-chlorophenyl-4-chloro-benzoate (Fig. 6) [1g].

CIDNP effects have also been studied during photolysis of dibenzoyl peroxide or substituted dibenzoyl peroxides in various solvents. [1f,1f,4b,7]. For solutions of the peroxides in CCl_4, CH_2Cl_2, methylformiate, monochloroacetic acid, in binary mixtures of these solvents and in CCl_4/CBr_4- and CCl_4/CH_3J mixtures, enhanced absorptions of the phenyl benzoates and emissions of the substituted benzenes were observed as in the thermal decomposition reactions. This indicates that photolysis of aroyl peroxides occurs from excited singlet states in these solvents. For CCl_4 solutions containing small amounts of anthracene, the emission signal of chlorobenzene was found to be larger than for pure CCl_4 solutions and this was taken as evidence for sensitization by excited singlet anthracene [4b]. A charge-transfer complex formation between ground-state benzoyl peroxide and excited singlet anthracene may be involved [7].

The benzene or substituted benzene emissions may change to enhanced absorptions and the phenylbenzoate absorptions may change to emissions if dibenzoyl peroxides are irradiated in solutions containing such added effective triplet sensitizers, as Michler's ketone, 2-acetonaphthone, 1-acetonaphthone, acetophenone, benzophenone, acetone and cyclohexanone [1f,4b,7]. An example is given in Fig. 7, which should be compared with Fig. 6. Assuming that dibenzoyl peroxide decomposes from an excited triplet state ($\mu > 0$), the change of polarizations with addition of triplet sensitizers is easily explained. With $\mu > 0$, Eq. (36) now reads

$$\Gamma_{ni} = + --+ = +$$

19

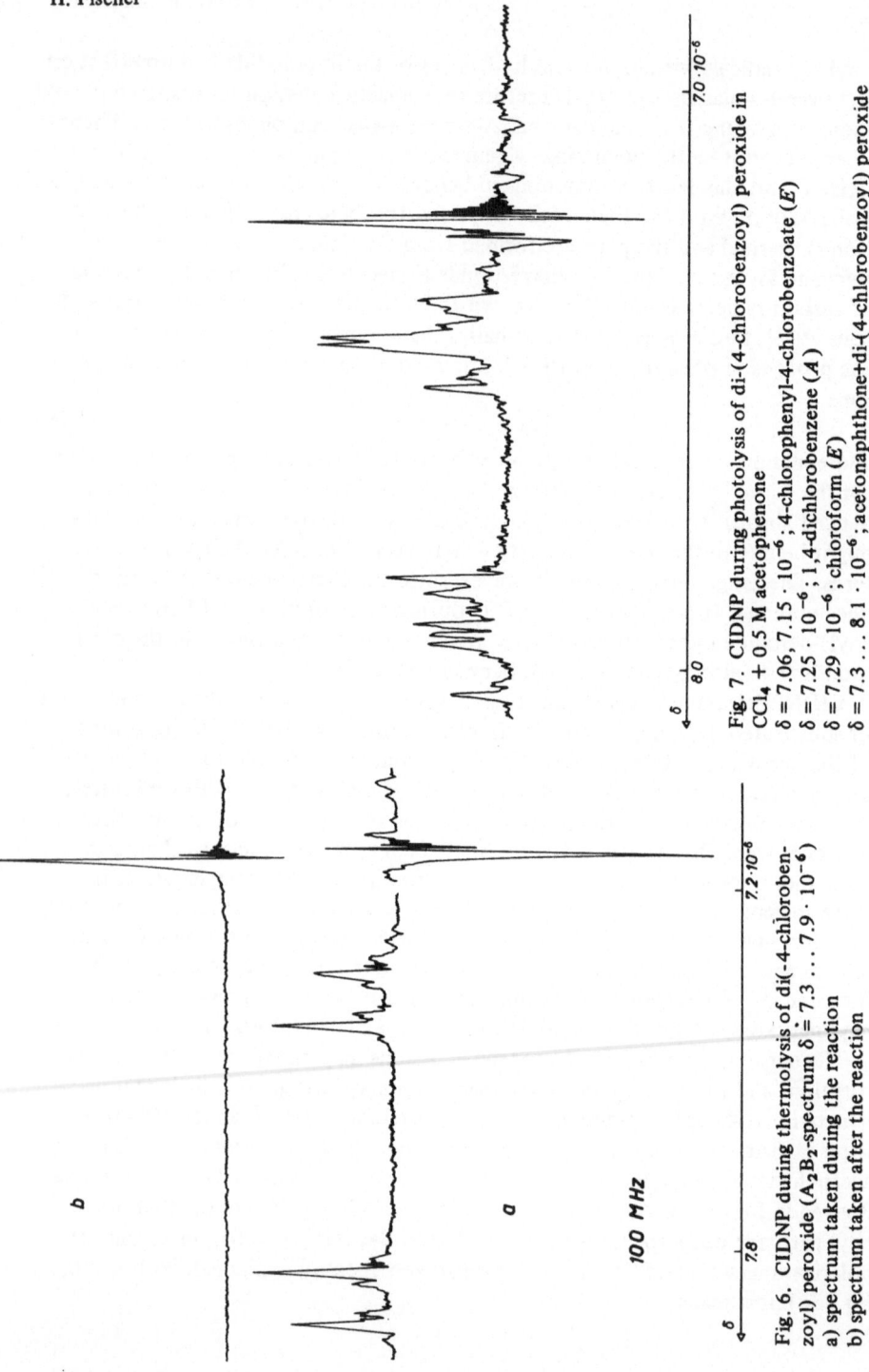

Fig. 7. CIDNP during photolysis of di-(4-chlorobenzoyl) peroxide in CCl$_4$ + 0.5 M acetophenone
δ = 7.06, 7.15 · 10^{-6}; 4-chlorophenyl-4-chlorobenzoate (E)
δ = 7.25 · 10^{-6}; 1,4-dichlorobenzene (A)
δ = 7.29 · 10^{-6}; chloroform (E)
δ = 7.3 ... 8.1 · 10^{-6}; acetonaphthone+di-(4-chlorobenzoyl) peroxide

Fig. 6. CIDNP during thermolysis of di(-4-chloroben-zoyl) peroxide (A_2B_2-spectrum δ = 7.3 ... 7.9 · 10^{-6})
a) spectrum taken during the reaction
b) spectrum taken after the reaction

100 MHz

for the benzenes and A is expected. In the sensitization experiments a change of polarizations with wavelength of irradiation was observed [1f,4b], indicating competition between direct (singlet) and sensitized (triplet) decompositions.

The changes of polarizations with the addition of various sensitizers may be regarded as strong evidence for the radical pair mechanisms of CIDNP and provide a nice example of the dependence of the effects on the precursor states. Two other important features of CIDNP also become apparent from the results of the studies outlined above.

The first ist the strong dependence of observed CIDNP intensities on the nuclear spin lattice relaxation times T_1 of the products. As is seen from Eqs. (25) and (8), the enhancement factor per formed molecule of a special transition $K \to L$ of the "escape" product benzene should be related to that of the corresponding transition of the "cage" product phenylbenzoate by

$$V_{KL}\ (e) = -\ \kappa \cdot V_{KL}\ (c) \cdot \frac{r(c)}{r(e)} \qquad (39)$$

κ allows for nuclear spin relaxation in the "escaping" radicals and contains reaction terms. (Sect. 4.3.) Since the relaxation diminishes level population differences, κ should be $0 \leqslant \kappa \leqslant 1$. Thus, the enhancement factor per benzene molecule formed should be smaller in absolute magnitude than that of the phenylbenzoate. In Figs 1, 6, and 7 the observed signals are clearly larger for the benzene than for the benzoate. Since the observed signals are proportional to the product relaxation times T_1 besides the V_{KL}, this might be due to grossly different relaxation times of the two products. In particular, insertion of (39), (8), and (5) into (3) gives

$$\frac{T_1(e)}{T_1(c)} \cdot \frac{I_{KL}(c) - I^0_{KL}(c)}{I_{KL}(e) - I^0_{KL}(e)} = -\frac{1}{\kappa} \qquad (40)$$

a ratio which should be smaller than -1. Under conditions corresponding to those used to obtain Fig. 6, $T_1 = 4.5$ sec for 4-Cl-phenyl-4-Cl-benzoate and $T_1 = 26.4$ sec for 1,4-dichlorobenzene[1g], and with the observed intensities it becomes -2. Thus (39) also holds in this case, despite the apparent behaviour. Obviously, the relaxation times of the products have to be considered carefully in quantitative discussions of observed signal intensities, and a product showing the largest signal need not necessarily be the one polarized in the pairs to the highest extent.

Secondly, it will be noted that quite a minor reaction pathway in Scheme VI leads to the CIDNP effects, since the "cage" product phenylbenzoate has a yield of only 4%. The major decomposition reaction (1) involving benzoyloxy/benzoyloxy pairs does not produce CIDNP effects. Therefore CIDNP effects, while giving evidence of free radicals, do not indicate that free radicals are the only intermediates in a chemical reaction.

21

4.2 CIDNP During Photolysis of Dialkyl Ketones

Strong CIDNP effects may be observed during photolysis of dialkyl ketones [21-23], and an example will be treated in this Section. Fig. 8a shows an NMR spectrum taken during irradiation of a 0.7 M solution of *tert.-butyl methyl ketone* (pinacolone) in benzene with the unfiltered light of a high-pressure mercury lamp [23]. When compared with the spectrum taken after 20 minutes' ir-

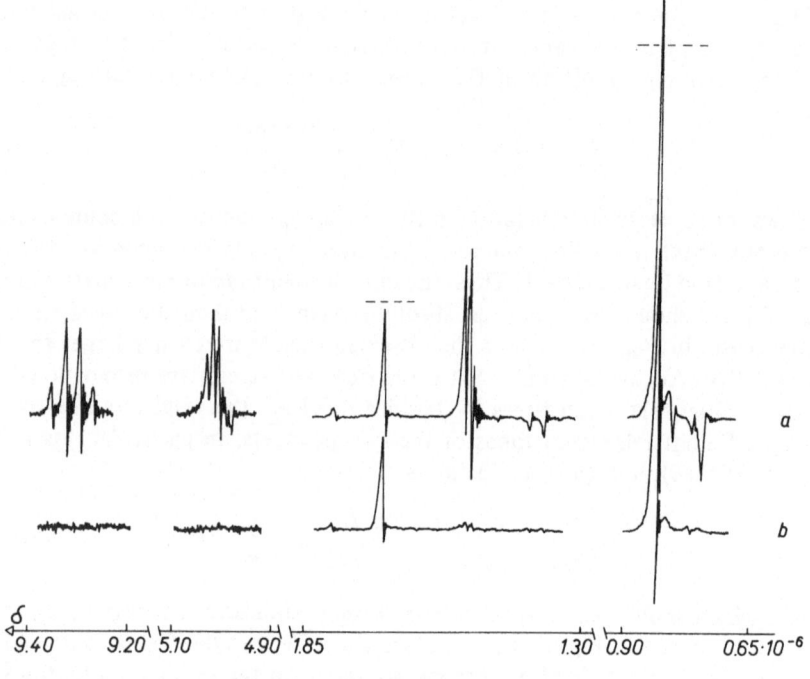

Fig. 8. CIDNP during photolysis of tert.-butyl-methyl ketone 0.7 M in benzene
a) spectrum taken during irradiation
b) spectrum taken after 20 minutes irradiation
$\delta = 0.84 \cdot 10^{-6}$; tert.-butyl group of the ketone (---- intensity before and after 90 sec irradiation)
$\delta = 1.70 \cdot 10^{-6}$; methyl group of the ketone (---- intensity before and after 90 sec irradiation)
$\delta = 1.40 \cdot 10^{-6}$; acetaldehyde ($CH_3$), doublet
$\delta = 9.20 \cdot 10^{-6}$; acetaldehyde (CHO), quartett
$\delta = 1.53 \cdot 10^{-6}$; isobutylene ($CH_3$), triplet
$\delta = 5.00 \cdot 10^{-6}$; isobutylene ($CH_2$), septett
$\delta = 0.78 \cdot 10^{-6}$; isobutane ($CH_3$), doublet
$\delta = 0.77 \cdot 10^{-6}$; hexamethylethane, $\delta = 1.80 \cdot 10^{-6}$ biacetyl

radiation (Fig. 8b), Fig. 8a demonstrates CIDNP effects for the products pinacolone, acetaldehyde, isobutylene, isobutane, biacetyl and hexamethylethane, which are listed in Table 1. The products are in agreement with results of photochemical studies on the same or similar systems[44,45]. Their formation may be rationalized as shown in Scheme VII, 1, and the radical pairs acetyl/tert.-butyl,

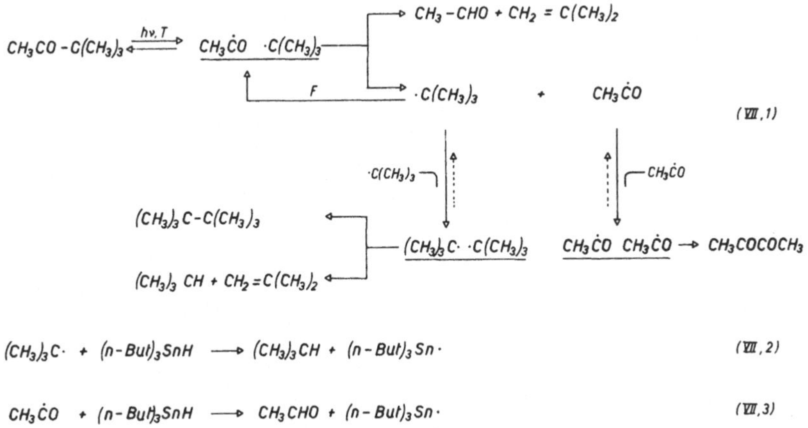

$$CH_3CO - C(CH_3)_3 \xrightarrow{h\nu, T} CH_3\dot{C}O \quad \cdot C(CH_3)_3$$

$$\longrightarrow CH_3 - CHO + CH_2 = C(CH_3)_2$$

$$\longrightarrow \cdot C(CH_3)_3 \quad + \quad CH_3\dot{C}O \quad (\mathit{VII}, 1)$$

$$(CH_3)_3 C - C(CH_3)_3 \longleftarrow$$

$$(CH_3)_3 CH + CH_2 = C(CH_3)_2 \longleftarrow$$

$$\longrightarrow (CH_3)_3 C \cdot \cdot C(CH_3)_3 \quad CH_3\dot{C}O \; CH_3\dot{C}O \longrightarrow CH_3COCOCH_3$$

$$(CH_3)_3C \cdot + (n - But)_3SnH \longrightarrow (CH_3)_3CH + (n - But)_3Sn \cdot \qquad (\mathit{VII}, 2)$$

$$CH_3\dot{C}O + (n - But)_3SnH \longrightarrow CH_3CHO + (n - But)_3Sn \cdot \qquad (\mathit{VII}, 3)$$

Scheme VII [23]

acetyl/acetyl and tert.-butyl/tert.-butyl are involved. According to Scheme VII, 1, the acetyl/tert.-butyl radical pairs are formed in the primary decomposition reaction of pinacolone (Norrish type I cleavage) and by random encounters of freely diffusing acetyl and tert.-butyl radicals which escaped the primary pairs. Acetyl/acetyl and tert.-butyl/tert.-butyl pairs are formed by random radical encounters only. Presumably, the Norrish type I cleavage occurs predominantly from an excited triplet state of pinacolone [44,45].

As is also shown in Table 1, all the observed CIDNP effects are easily explained with this assumption, with Scheme VII, 1, and Eqs. (36) and (37), if the following set of parameters is chosen

$$a_H (CH_3 \dot{C}O) = + 5.1 \, G$$

$$g (CH_3 \dot{C}O) = 2.0007$$

$$a_H \{(CH_3)_3 \dot{C}\} = + 22.74 \, G$$

$$g \{(CH_3)_3 \dot{C}\} = 2.0025$$

$$J (CH_2 = C (CH_3)_2) < 0$$

Apart from the positive sign of $a_H (CH_3 \dot{C}O)$, these parameters are known [42,46,47]. Thus the agreement between calculated and oberved polarizations may be considered to support the assumptions.

Further support for Scheme VII, 1 is obtained from experiments studying the influence on the polarizations during photolysis of benzene solutions of

Table 1. *CIDNP During Photolysis of Pinacolone in Benzene*

Product	Polarization exptl.		Pairs involved	Prec.	Product type	Γ_n	Γ_m	Polarization calc.
$CH_3CO—C(CH_3)_3$	$\underline{CH_3}\,CO$	E	$CH_3\dot{C}O \cdot C(CH_3)_3$	T,F	c	++-+	0	E
	$C(\underline{CH_3})_3$	A	"	"	"	++++	0	A
CH_3CHO	$\underline{CH_3}$	$E+AE$	"	"	"	++-+	+++++-	$E+AE$
	$\underline{CHO}$	$A+AE$	"	"	"	++++	+++++-	$A+AE$
$CH_2 = C(CH_3)_2$	$\underline{CH_2}$	$A+AE$	"	"	"	++++	++++-+	$\left.\begin{array}{c} \\ \end{array}\right\} A+AE$
			$+(CH_3)_3C\cdot \ \cdot C(CH_3)_3$		"	0	+++-+	
	$\underline{CH_3}$	$A+AE$	$CH_3\dot{C}O \cdot C(CH_3)_3$	T,F	"	++++	+++++-	$\left.\begin{array}{c} \\ \end{array}\right\} A+AE$
			$+(CH_3)_3C\cdot \ \cdot C(CH_3)_3$	F	"	0	++++-+	
$CH_3CO—OCCH_3$	$\underline{CH_3}$	A	$CH_3\dot{C}O \cdot C(CH_3)_3$	T,F	e	+--+	0	A
			$+CH_3\dot{C}O \ OCCH_3$	F	c	0	0	
$(CH_3)_3C—C(CH_3)_3$	$\underline{CH_3}$	E	$CH_3\dot{C}O \cdot C(CH_3)_3$	T,F	e	+-++	0	$\left.\begin{array}{c} \\ \end{array}\right\} E$
			$+(CH_3)_3C\cdot \ \cdot C(CH_3)_3$	F	c	0	0	
$(CH_3)_3\,CH$	$\underline{CH_3}$	$E+AE$	$CH_3\dot{C}O \cdot C(CH_3)_3$	T,F	e	+-++	+++++-	$\left.\begin{array}{c} \\ \end{array}\right\} E+AE$
			$+(CH_3)_3C\cdot \ \cdot C(CH_3)_3$	F	c	0	+++++-	

pinacolone of adding the effective radical scavenger tri-n-butyl tin hydride. Reactions VII,2 and VII,3 should compete with pair formations by random encounters, and this competition leads to a decrease of the polarizations of pinacolone, acetaldehyde, isobutylene, biacetyl and hexamethylethane and to a change in isobutane polarization from mainly multiplet type AE (Fig. 8) to mainly net effect E.

On addition of the triplet quenchers stilbene or piperylene to solutions of pinacolone, all polarizations strongly decrease. Low piperylene concentrations, in particular, suppress the isobutane polarizations completely while they affect those of acetaldehyde and isobutylene to a lesser extent. Piperylene may also act as radical trap. For 7 M solutions in piperylene, very weak polarizations of isobutylene and acetaldehyde of the types given in table 1 can still be observed. Apparently the polarizations may arise from primary triplet pairs in this case, too[45].

CIDNP effects similar to those described here are found with other acyclic aliphatic ketones in benzene or n-hexane solutions [21-23]. They can be explained in terms of reactions like those of Scheme VII,1. If CCl_4 or other chlorinated solvents are applied, primary pair formation from singlet precursors seems possible [22-23].

4.3 Comparisons of CIDNP of "Cage" Combination or Disproportionation and "Escape" Products

One of the fundamental results of the radical pair theory of CIDNP is that it relates the polarizations of products formed by direct combination or disproportionation of the polarizing geminate radical pairs with that of products derived from transfer or coupling reactions of radicals escaping these pairs (Sect. 2, rule A; Sect. 3, Eq. (25, 27, 36, 37); Sect. 4.1, Eq. (39)). Quantitatively, this relation is given by equations

$$p_k(e) = \kappa \cdot \frac{1}{r(e)} \, (r_k - r(c) \cdot p_k(c)) \tag{25}$$

and

$$\underline{V_{KL}}(e) = - \kappa \cdot \underline{V_{KL}}(c) \cdot \frac{r(c)}{r(e)} \tag{39}$$

where $0 \leqslant \kappa \leqslant 1$ allows for reactions and nuclear relaxation of the "escaping" radicals. On the basis of simple kinetic considerations, an analytic form of κ can be derived [1f,6f]. If we denote the rate of formation of a "cage" product with nucleus i of R in spin state k by $r_k(c)$, then the rate of formation of a free radical R_i with nucleus i in the same spin state k will be given by

$$r'_k = r_k - r_k(c) \tag{41}$$

where r_k is the total rate of pair formation with i in level k. The rate equation for the concentration of R_k is then

$$\frac{d[R_k]}{dt} = r'_k - \frac{[R_k] - [R_k^0 \cdot]}{T_{1R}} - F \cdot [R_k]$$ (42)

where $[R_k^0 \cdot]$ is the thermal equilibrium concentration of $R \cdot$ in level k, T_{1R} is the nuclear relaxation time of i in $R \cdot$ and F is a factor which is determined by the reactions of $R \cdot$ giving the escape products. If the radicals $R \cdot$ are trapped by a suitable scavenger SX and vanish by transfer reactions,

$$F = k_s \cdot [SX]$$ (43)

where k_s is the rate constant of scavenging. If they react in consecutive radical-radical reactions,

$$F = k_c [R\cdot]$$ (44)

where $[R\cdot]$ is the total radical concentration. If $[R_k \cdot] \gg [R_k^0 \cdot]$, (42) can be easily solved for $[R_k \cdot]$. Under steady-state conditions from (42, 41, 8, 9)

$$\underline{V_{KL}}(e) = -\frac{F}{F + \dfrac{1}{T_{1R}}} \cdot \underline{V_{KL}}(c) \cdot \frac{r(c)}{r(e)}$$ (45)

defining κ, is obtained.

It will be noted that the nuclear relaxation in $R \cdot$ is described by a single relaxation time T_{1R}. This is a simplification and the reservations outlined in Sect. 3.1. apply again. However, two quantitative comparisons of the CIDNP of "cage" and "escape" products have been published [1f,6f] and (45) was found to hold to a very good approximation.

In [1f] the polarizations of the products of the reactions given in Scheme V and their dependence on solvent composition $[CH_3 CO CH_3]/[CH_2 Cl_2]$ were determined. The ratios of the intensities of the $CHCl_2$ resonances of $CHCl_2-CHCl_2$ (I_s) and $CHCl_2-CH_2COCH_3$ (I_A) were found to depend on the solvent composition and on the total rates of radical pair production r_{tot}. Denoting the rate constants of the reactions

$$\phi\cdot + CH_2 Cl_2 \quad \xrightarrow{k_1} \quad \phi + \cdot CHCl_2$$
$$\phi\cdot + CH_3 CO CH_3 \quad \xrightarrow{k_2} \quad \phi + \cdot CH_2 CO CH_3$$ (46)
$$\cdot CHCl_2 + \cdot CHCl_2 \quad \xrightarrow{k_3} \quad CHCl_2 - CHCl_2$$

as k_1, k_2, and k_3, we obtain from (45), (44), (7), and (5), and $I \gg I_0$

$$\frac{I_A}{I_S} = -\frac{T_{1A}}{T_{1S}} \left\{ 1 + \frac{1}{T_{1R}} \sqrt{\frac{2}{k_3 \cdot r_{tot}}} \cdot x \right\}$$ (47)

where

$$x = 1 + \frac{k_2}{k_1} \frac{[CH_3COCH_3]}{[CH_2Cl_2]} \qquad (48)$$

for these dependences. In Fig. 9 the experimental results are seen to agree with the theoretical predictions: From the slopes of the straight lines of Fig. 9 a value for the nuclear relaxation time T_{1R} of $\cdot CHCl_2$ is deduced as $T_{1R} = 4.5 \cdot 10^{-4}$ sec. In a similar study, Closs[6f] has found $T_{1R} = 3.5 \cdot 10^{-4}$ sec for another radical.

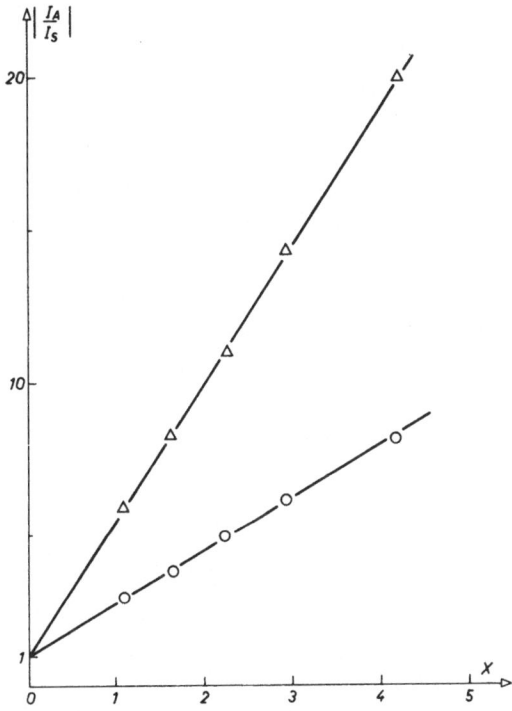

Fig. 9. Ratios of $CHCl_2$-transition intensities of $CHCl_2-CHCl_2$ (I_S) and $CHCl_2-CH_2COCH_3$ (I_A) during photolysis of dibenzoyl peroxide in CH_2Cl_2/CH_3COCH_3 mixtures (100 MHz, $\Delta : r_{tot} = 10^{-4}$ M $\cdot$ sec^{-1}, O : $r_{tot} = 6 \cdot 10^{-4}$ M sec^{-1})

4.4 Magnetic Properties of Free Radicals as Determined by CIDNP

The signs and amplitudes of nuclear polarizations depend on chemical parameters and on the hyperfine coupling constants and g factors of the free radicals involved in the reactions. If the chemical parameters are known, the radical properties can be determined from the CIDNP patterns. Even if the chemical parameters are not known, CIDNP patterns usually contain enough information to allow simultaneous determinations of all the parameters. Thus, CIDNP may be used like electron spin resonance (ESR) as a tool to obtain magnetic properties of free radicals. Since CIDNP effects depend on the signs of hyperfine coupling constants as well as on their magnitudes, CIDNP might even be superior to ESR in some cases.

Only a few applications of CIDNP with regard to the determination of radical properties have been published so far. Chapter 4.2. reports that a positive sign of a_H (CH$_3$ĊO) was found to agree with the observed CIDNP effects, whereas only the absolute value of this quantity was known from ESR studies. In Ref. [1f] the parameters

$$a_H (\cdot CHCl_2) = (-17.0 \pm 1.0) \text{ gauss}$$

$$g (\cdot CHCl_2) = 2.0080 \pm 0.003$$

were determined. This radical has not so far been detected by ESR. For the radical pairs $R\cdot$ $\cdot R'$

$$p - x \, C_6H_4\dot{C}HOH \quad \dot{C}H (C_6H_4\text{-p-y})_2$$

a: x = Br; y = H
b: x = Cl; y = H
c: x = H; y = H
d: x = H; y = Cl
e: x = H; y = Br

the g factor differences $g - g'$ were taken from CIDNP patterns by Closs [6d] and by Adrian [32]. These authors applied different theoretical formulae and obtained slightly different values, [6d]: a: $2.7 \cdot 10^{-3}$; b: $1.5 \cdot 10^{-3}$; c: $0.47 \cdot 10^{-3}$; d: $-0.33 \cdot 10^{-3}$; e: $-2.7 \cdot 10^{-3}$; [32]: a: $1.57 \cdot 10^{-3}$; b: $1.24 \cdot 10^{-3}$; c: $0.72 \cdot 10^{-3}$; d: $-0.08 \cdot 10^{-3}$; e: $-1.78 \cdot 10^{-3}$, but the trends of the g factor differences are well represented by both results.

Further CIDNP studies on radical properties may be expected to reveal the importance of the new technique as a powerful complement to ESR.

5. Concluding Remarks

The examples given in the preceding section have shown the potentialities of CIDNP as a tool for the determination of reaction mechanisms and radical properties; they should also demonstrate that CIDNP does provide the various data listed in the introduction in rather straightforward procedures.

Here we wish to emphazise a few points which make clear that the method has its limitations and to indicate that the theory is still far from fully developed.

In the dibenzoylperoxide decompositions the effects are caused in pairs which are themselves part of a minor reaction pathway. The same may be true for other reactions, especially of the rearrangement type [14-16], where ionic and radical intermediates may be present simultaneously. CIDNP effects are evidence for radical intermediates, though others may be present as well, and for quantitative studies CIDNP has to be combined with other techniques to elucidate the relative importance of various possible pathways.

As will be evident from Sect. 3.1., the amplitudes of the effects are determined by a variety of factors. In particular, Eq. (8) relates the observed enhancements to the relaxation times of the products, the rates of product formation and the enhancement factors per product molecule formed. From experience [1f,g,4,6], the latter might be expected to be of the order of 100 to 1000 for protons and high fields. Therefore CIDNP effects may be anticipated ($V_{KL} \geqslant 10$) when $r/n \cdot T_1 \leqslant 10^{-2} \ldots 10^{-1}$. Since T_1 is normally of the order of 1 to 10 sec for protons, a reaction to be studied by CIDNP should proceed rather rapidly. No general rule can be given, but from the above considerations a characteristic time constant of the reaction, for instance, the time for 50 % conversion, of about 10 to 20 min or less is advisable. This requirement can often be met by the application of suitable reaction conditions.

Finally, it should be noted that the effects are field-dependent. In one case [11] no effects were seen when the reactions were run within the spectrometer fields, but large effects were observed for samples which were allowed to react in the earth's field or in a low field of a separate magnet and which were transferred to the probe of the spectrometer shortly after the end of the reaction. These field effects on CIDNP will deserve much future investigation. So far, only a limited amount of information is available [1e,2f,11,39,48,49] and no thorough theoretical study has been reported. Preliminary results indicate, however, that the low-field polarizations are explainable in terms of suitable extensions of the high-field theories of Sect. 3.2. [33,39,48,49]. Future investigations have also to establish the potentialities of CIDNP in the detection of biradical intermediates. Recently Closs [6k] reported CIDNP effects for reactions where such intermediates are present and outlined the pertinent theory which indicates severe restrictions.

Summarizing, it may be stated that the exploitation of the relatively young technique of CIDNP may be expected to proceed in the following directions: on the theoretical side, a general theory based on first principles will be developed, covering amplitudes and signs of low- and high-field effects and biradical intermediates without resorting to experimental parameters; on the experimental side, the list of reactions studied, mechanisms elucidated and radical parameters determined by CIDNP will expand rapidly.

H. Fischer

It is a pleasure to acknowledge stimulating discussions and correspondence with Drs. Adrian, Closs, Glarum, Kaptein, Lawler and Ward and the communication of their results prior to publication. I wish to thank especially B. Blank and M. Lehnig of this laboratory for providing most of the experimental results presented in this report.

References

[1a] Bargon, J., Fischer, H., Johnson, U.: Z. Naturforsch. *22a*, 1551 (1967).
[1b] − − Z. Naturforsch. *22a*, 1556 (1967).
[1c] − − Z. Naturforsch. *23a*, 2109 (1968).
[1d] − − Acc. Chem. Res. *2*, 110 (1969).
[1e] Lehnig, M., Fischer, H.: Z. Naturforsch. *24a*, 1771 (1969).
[1f] − − Z. Naturforsch. *25a*, 1963 (1970).
[1g] Blank, B., Fischer, H.: Helv. chim. Acta *54*, 905 (1971).
[2a] Ward, H. R., Lawler, R. G.: J. Am. Chem. Soc. *89*, 5518 (1967).
[2b] Lawler, R. G.: J. Am. Chem. Soc. *89*, 5519 (1967).
[2c] Ward, H. R., Lawler, R. G., Loken, H. Y.: J. Am. Chem. Soc. *90*, 7359 (1968).
[2d] − − Cooper, R. A.: J. Am. Chem. Soc. *91*, 746 (1969).
[2e] − − − Tetrahedron Letters *1969*, 527.
[2f] − − Loken, H. Y., Cooper, R. A.: J. Am. Chem. Soc. *91*, 4928 (1929).
[2g] − − Marzilli, T. A.: Tetrahedron Letters *1970*, 521.
[2h] − Acc. Chem. Res., in press.
[2i] − paper presented at the International Colloquium on CIDNP, Brussels, 18.−19.3.1971.
[3] Hausser, K.-H., Stehlik, D.: Advan. Magn. Resonance *3*, 79 (1968).
[4a] Kaptein, R.: Chem. Phys. Lett. *2*, 261 (1968).
[4b] − Hollander, J. A. den, Antheunis, D., Oosterhoff, L. J.: Chem. Commun. *1970*, 1687.
[4c] − Verhuis, F. W., Oosterhoff, L. J.: in press.
[4d] − Frater-Schröder, M., Oosterhoff, L. J.: in press.
[5a] Rykov, S. V., Buchachenko, A. L.: Dokl Akad. Nauk SSSR *185*, 870 (1969).
[5b] − − Dodonov, V. A., Kessenich, A. V., Razuvaev, G. A.: Dokl. Akad. Nauk SSSR *189*, 341 (1969).
[5c] − − Baldin, V. I.: J. Strukt. Chem. (SSSR) *10*, 814 (1969).
[5d] Buchachenko, A. L., Rykov, S. V., Kessenich, A. V.: J. Phys. Chem. (Moscow) *44*, 488 (1970).
[5e] − − − Bylina, G. S.: Dokl. Akad. Nauk SSSR *190*, 839 (1970).
[5f] Lippmaa, E., Pehk, T., Buchachenko, A. L., Rykov, S. V.: Chem. Phys. Lett. *5*, 521 (1970).
[5g] − − Saluveere, S., paper presented at the International Colloqium on CIDNP, Brussels, 18.−19.3.1971.
[5h] Rykov, S., Buchachenko, A. L., Kessenich, A. V.: Spectr. Letters *3*, 55 (1970).
[6a] Closs, G. L., Closs, L. E.: J. Am. Chem. Soc. *91*, 4549 (1969).
[6b] − − J. Am. Chem. Soc. *91*, 4550 (1969).
[6c] − Trifunac, A. D.: J. Am. Chem. Soc. *91*, 4554 (1969).

6d) Closs, G. L., Doubleday, C. E., Paulson, R.: J. Am. Chem. Soc. *92*, 2185 (1970).
6e) – Trifunac, A. D.: J. Am. Chem. Soc. *92*, 2186 (1970).
6f) – – J. Am. Chem. Soc. *92*, 2186 (1970).
6g) – Paulson, D. R.: J. Am. Chem. Soc. *92*, 7229 (1970).
6h) – paper presented at the International Colloquium on CIDNP, Brussels, 18.–19.3.1971.
7) Fahrenholtz, S. R., Trozzolo, A. M.: in press.
8a) Lepley, A. R.: J. Am. Chem. Soc. *90*, 2710 (1968).
8b) – Chem. Commun. *1969*, 64.
8c) – Landau, R. C.: J. Am. Chem. Soc. *91*, 748 (1969).
8d) – J. Am. Chem. Soc. *91*, 749 (1969).
8e) – J. Am. Chem. Soc. *91*, 1237 (1969).
8f) – Cook, P. M., Willard, G. F.: J. Am. Chem. Soc. *92*, 1101 (1970).
8g) Walling, Ch., Lepley, A. R.: J. Am. Chem. Soc. *93*, 546 (1971).
9) Koenig, T., Mabey, W.: J. Am. Chem. Soc. *91*, 1237 (1969).
10) Rashkys, J. W.: Chem. Commun. *1970*, 578.
11a) Garst, J. F., Cox, R. H., Barbas, J. T., Roberts, R. D., Morris, J. I., Morrison, R. C.: J. Am. Chem. Soc. *92*, 5761 (1970).
11b) – – J. Am. Chem. Soc. *92*, 6389 (1970).
12) Lane, A. G., Rüchardt, Werner, R.: Tetrahedron Letters *1969*, 3213.
13) Rieker, A., Niederer, P., Leibfritz, D.: Tetrahedron Letters *1969*, 4287, and paper presented at the International Colloquium on CIDNP, Brussels, 18 - 19.3.1971.
14a) Schöllkopf, U., Ostermann, G., Schossig, J.: Tetrahedron Letters *1969*, 2619.
14b) – Ludwig, U., Ostermann, G., Patsch, M.: Tetrahedron Letters *1969*, 3415
14c) – Hoppe, I.: Tetrahedron Letters *1970*, 4527.
14d) – Angew. Chemie *82*, 795 (1970).
14e) – paper presented at the International Colloquium on CIDNP, Brussels, 18.–19.3.1971.
15a) Jemison, R. W., Morris, D. G.: Chem. Commun. *1969*, 1226.
15b) Morris, D. G.: Chem. Commun. *1969*, 1345.
15c) – paper presented at the International Colloquium on CIDNP, Brussels, 18.–19.3.1971.
16a) Baldwin, J. E., Brown, J. E.: J. Am. Chem. Soc. *91*, 3647 (1969).
16b) – Erickson, W. E., Hackler, R. E., Scott, R. M.: Chem. Commun. *1970*, 576.
16c) – Brown, J. E., Höfle, G.: J. Am. Chem. Soc. *93*, 788 (1971).
17a) Iwamura, H., Iwamura, M., Nishida, T., Yoshida, M., Nakayama, J.: Tetrahedron Letters *1971*, 63.
17b) – – – Sato, S.: Bull. Chem. Soc. Japan *43*, 3638 (1970).
17c) – – – – J. Am. Chem. Soc. *92*, 7474 (1970).
17d) – Tetrahedron Letters *1970*, 3723.
18) Holländer, J., Neumann, W. P.: Angew. Chemie *82*, 813 (1970), and paper presented at the International Colloquium on CIDNP, Brussels, 18.–19.3.1971.
19a) Cocivera, M., Roth, H. D.: J. Am. Chem. Soc. *92*, 2573 (1970).
19b) Roth, H. D.: Paper presented at the International Colloquium on CIDNP, Brussels, 18.–19.3.1971.
20) Cocivera, M., Trozzolo, A. M.: J. Am. Chem. Soc. *92*, 1774 (1970).
21) DoMinh, T., Trozzolo, A. M.: Paper presented at the International Colloquium on CIDNP, Brussels, 18 - 19.3.1971.
22) Hollander, J. A., den: Paper presented at the International Colloquium on Brussels, 18 - 19.3.1971.
23) Blank, B., Mennitt, P. G., Fischer, H.: Unpublished.
24) Janzen, E. G.: Acc. Chem. Res. *4*, 31 (1971).
25) Lebin, Ya. A., Iljasov, A. V., Pobedinskij, D. G., Samupov, Yu. Yu., Sauda, I. I., Goldfars, E. I.: Isv. Acad. Nauk SSR (1970).

[26a] Fischer, H.: J. Phys. Chem. *73*, 3834 (1969).
[26b] Russell, G. A., Lamson, D. W.: J. Am. Chem. Soc. *91*, 3967 (1969).
[27] Cocivera, M.: J. Am. Chem. Soc. *90*, 3261 (1968).
[28] Abragam, A.: Principles of Nuclear Magnetism. Oxford: Clarendon Press 1961.
[29a] Closs, G. L.: J. Am. Chem. Soc. *91*, 4552 (1969).
[29b] — Trifunac, A. D.: J. Am. Chem. Soc. *92*, 2183 (1970).
[30] Kaptein, R., Oosterhoff, L. J.: Chem. Phys. Letters *4*, 195, 214 (1969).
[31a] Fischer, H.: Chem. Phys. Letters *4*, 611 (1970).
[31b] — Z. Naturforsch. *25a*, 1957 (1970).
[32a] Adrian, F. J.: J. Chem. Phys. *53*, 3347 (1970).
[32b] — Private communication.
[33] Glarum, S. H.: Private communication.
[34] Kaptein, R.: Private communication.
[35] Carrington, A., McLachlan, A.D.: Introduction to Magnetic Resonance. New York: Harper and Row 1967.
[36a] Noyes, R. M.: J. Chem. Phys. *22*, 1349 (1954).
[36b] — J. Am. Chem. Soc. *77*, 2042 (1955).
[36c] — J. Am. Chem. Soc. *78*, 5486 (1956).
[37] Dobis, O., Pearson, J. M., Szwarc, M.: J. Am. Chem. Soc. *90*, 278, 283 (1968).
[38] Ward, H. R., Lawler, R. G.: Accounts Chem. Res., in press.
[39] Kaptein, R.: Private communication.
[40] Lehnig, M., Fischer, H.: Unpublished.
[41a] Hammond, G. S., Soffer, L. M.: J. Am. Chem. Soc. *72*, 4711 (1950).
[41b] Swain, C. G., Schaad, L. J., Kresge, A. G.: J. Am. Chem. Soc. *80*, 5313 (1958).
[41c] Martin, J. C., Hargis, J. H.: J. Am. Chem. Soc. *91*, 5399 (1969).
[42] Fischer, H.: Magnetic Properties of Free Radicals. In: Landolt-Börnstein, New Series, Group II: Atomic and Molecular Physics (eds. K. H. Hellwege and A. M. Hellwege), Vol. 1. Berlin-Heidelberg-New York: Springer-Verlag 1965.
[43] Kasai, P. H., Clark, P. A., Whipple, E. B.: J. Am. Chem. Soc. *92*, 2640 (1970).
[44a] Wagner, P. J., Hammond, G. S.: Advan. Photochem. *5*, 21 (1968).
[44b] Dalton, J. C., Turro, N. J.: Ann. Rev. Phys. Chem. *21*, 499 (1970).
[45] Yang, N. C., Feit, D. E.: J. Am. Chem. Soc. *90*, 504 (1968).
[46] Bennett, J. E., Mile, B., Ward, B.: Chem. Commun. *1969*, 13.
[47] Emsley, J. W., Feeney, J., Sutcliffe, L. H.: High Resolution Nuclear Magnetic Resonance, Vol. 2. Oxford: Pergamon Press 1966.
[48a] Fischer, H., Lehnig, M.: J. Phys. Chem., in press.
[48b] Lehnig, M.: Private communication.
[49] Charlton, J. L., Bargon, J.: Chem. Phys. Letters *8*, 442 (1971).

Received May 6, 1971

Critique of the Notion of Aromaticity

Dr. Jean-François Labarre and Dr. François Crasnier

Départment de Chimie Inorganique, de l'Université Paul Sabatier, Toulouse (France)

The problem of Aromaticity* has always been one of the most difficult and fascinating problems in chemistry. It is usually stated in the literature that aromaticity has constituted a challenge to both the theoretican and the chemist since 1865, when August Kekule introduced his intuitive idea on the structure of the benzene molecule [1].

But the trouble is that the right question has never been asked. Which question have we to answer: "Is aromaticity an out-of-date or an up-to-date concept?" or (that most tedious question) "Is aromaticity a myth or reality?". These questions must be discussed before we can venture upon defining and measuring aromaticity.

It is commonly asserted that the words *"aromatic compounds"* were introduced at the end of the eighteenth century to distinguish such molecules from their aliphatic homologues because of their pleasant olfactory properties: natural products such as oil of wintergreen, aniseed, sassafras, oil of cinnamon and vanilla beans were included at this time in the *"aromatic class"* of compounds.

Here, then, is the first difficulty about the problem we are trying to solve: a scientist who discovers — as Kekule did — a fundamental new concept must surely introduce a *new an well-chosen word* to define it. If he does not do so, the risk is that complete confusion very soon occurs both in books and in the minds of men. We are of the opinion that one cause of the confusion which reigns today concerning the question: "What is aromaticity?" lies in Kekule's original choice of word, but it is also clear that the chemists who have worked

---

*This contribution was strongly influenced by the main ideas which were debated at the 3rd International Symposium on "Aromaticity, Pseudo-Aromaticity, Anti-Aromaticity" held in *Jerusalem* in March 1970.

in the field of aromaticity since 1865 have continued in the same frame of mind and have too often closed their eyes to what was being measured and defined when they spoke about aromaticity.

A second — and undoubtedly more specific — difficulty proceeds from the fact that aromaticity has at least two meanings which are fundamentally different according to whether the word is being used by a pure chemist or a physicist: to the first, a compound is aromatic if *its chemistry is like that of benzene,* while, to the other, according to the more modern definition, a compound is aromatic if it has *a low ground-state enthalpy.* [2]

Originally, the concept of aromaticity was developed as a means of characterizing a certain type of *organic* molecules which were inclined to substitution and disinclined to addition reactions and thermally stable [3]. The emphasis was for many years more upon the chemical activity than upon the physical properties in the ground state. The regenerative — or *meneidic,* according to Lloyd's proposal [4] — character of aromatic molecules and the fact that such a molecule must, as Kekule himself said [5], contain at least six carbon atoms explain why, ever since 1865, the problems concerning aromaticity were studies for some seventy years in exclusively *organic chemistry with benzene as a standard.* This helps us to understand why today it is surprising for organic chemists to hear inorganic chemists using their *hallowed word.*

In such conditions, it will be appreciated that, at the end of the nineteenth century and even at the beginning of the twentieth, aromaticity seemed a well-established and well-defined concept.

But, at the end of this happy time, came the advent of quantum chemistry which, it was hoped, would bring a comfortable support to what had gone before. Quantitative, increasingly accurate valence-bond and molecular-orbital methods were developed, permitting the calculation of the *resonance energy* of a conjugated system, a quantity which is a ground state property and can be measured experimentally and compared to the thermochemical data [6]. This was the start of a continuous process of transforming the meaning of aromaticity from the chemical definition — which emphasizes the energy content of the molecule in the *excited state* — to the physical viewpoint which underlines the properties of the molecules in the *ground state.* But the main consequence, which springs quite immediately out of the comparison between the chemical and the "new" physical definition of aromaticity, is that a large number of compounds are thus *physically aromatic* but *chemically not.* The cyclopentadienyl anion $C_5H_5^-$, prepared by Thiele in 1900 [7], but not recognized as such, is a good example of a non-regenerative molecule with respect to its high chemical reactivity which was found to possess a particularly low enthalpy in the ground state Consequently, the class of non-benzenoid compounds appears in the literature to designate mainly the "physically aromatic" molecules [8].

This was the beginning of the undermining of the word aromaticity and, from this time, each scientist concerned with related problems introduced into the breach that had been opened a new word for his own purposes (*new but always derived from aromaticity*). The mushrooming – as Bergmann and Agranat say so felicitously [9] – of prefixes was beginning. There are:

1 *pseudo*-aromaticity [10,11,12]
2 *quasi*-aromaticity [13]
3 *anti*-aromaticity [14,15]
4 *non*-aromaticity [14]
5 *homo*-aromaticity [16,17]
6 *pseudo-anti*-aromaticity [18]

and perhaps even *super*-aromaticity, *spiro*-aromaticity etc.

These concepts were introduced in order to bridge the gap between the two definitions of aromaticity mentioned above, but it is clear that the main result of this attempt has, in fact, been complete confusion.

Moreover, it is quite certain that this new tendency has something to do with the trouble into which many people plunged when they had to compare the aromaticity of any organic or *inorganic* molecule to that of benzene. This difficulty became more acute, simultaneously with the synthesis and increasing knowledge of the chemical and physical behaviour of *inorganic ring systems,* and these cases soon made it necessary to propose new definitions of the concept of aromaticity which would eliminate the need to refer systematically to benzene or benzenoid compounds.

Before developing this viewpoint, we wish to discuss and criticize briefly the "typically organic concept" we have already mentioned.

Let us first of all analyse Hückel's famous $(4n + 2)$ aromaticity rule [19,20] which played a leading role in the theory of non-benzenoid compounds. This rule was derived from theoretical considerations, but it should be pointed out that, despite its success in explaining and predicting many experimental facts [21,22], there exists, even for monocyclic conjugated hydrocarbons, an obvious discrepancy between theory and experiment. Indeed, the H. M. O. calculations show that if a $4n$-electron system has a lower resonance energy per atom than the homologous $(4n + 2)$ molecule, the magnitude of the difference is much less than experiment would suggest. Despite the improvements introduced by Coulson and Longuet-Higgins [23,24] and generalized by Dewar [25], who applied Hückel's rule extensively to the problem of aromaticity [26,27], the theoretical foundations of this rule gradually became weakened, particularly with the introduction of the bond alternation concept [28,29].

Many attemps were made to broaden Hückel's rule [30,31,32], mainly in the domain of *anti-aromatic* systems [33], but it quickly became evident that the only theoretical way to support a reasonable use of this rule to predict reality was to perform quantum calculations which would explicitly take into account the effects of interelectronic and internuclear repulsions and other terms generally neglected in the empirical methods. One interesting attempt, based on

Pople's method [34, 35], was made by Fukui and his co-workers [36]; this permitted progress in the understanding of the "push-pull" synthesis [37, 38, 39, 40]. Various other attempts were based on S.C.F.-M.O. calculations by Nakajima [41, 42, 43], Julg [44] and Daudel [45]. But it is most noticeable that all this work was done in the field of organic chemistry and never until today, on inorganic ring systems. This may be due to the lack of interest shown by too many quantum chemists, until the last two decades, in "aromatic" inorganic compounds; or it may be due to the fact that Hückel's rule is extremely difficult to apply in such cases, particularly since the calculations *must* be done in inorganic chemistry only by way of very accurate non-empirical methods.

The attempts of Dewar and co-workers [25, 46, 47, 48] to define Hückel's rule as a criterion of aromaticity by applying Mulliken and Parr's definition of resonance energy may be criticized on the same grounds. It became evident, following the publication of Dewar's famous 1952 paper, that resonance energy was a suitable indicator of the aromatic character of an organic molecule, and much use was made of this criterion by organic chemists. But, once more, the trouble affected "the rest": in the field of *inorganic* ring systems, the use of pure semiempirical calculations (which allow the resonance energy of organic molecules to be estimated correctly) is quite strictly forbidden on account of the heteronuclearity of the σ skeleton. Only overall electron methods may be used and a mixing of σ and π levels often occurs in such computations. The separation between these two types of levels, which is classically the foundation of the calculation of resonance energy in organic chemistry, vanishes in inorganic chemistry. Let us give a proof for this assertion: contrary to the results of "organic-like" calculations made on the molecule of borazole [49], the overall electron technics used by Brown and MacCormack [50] and by Davies [51], for example, allow these authors to conclude that the highest occupied molecular level is of σ and not π type. The complexity of the problem is furthermore underlined by the recent results obtained with overall electron methods: some of them give a highest σ level, others a π level, and there is no convenient explanation of this discrepancy.

Moreover, the calculation of resonance energy requires a knowledge of standard energies, that is, the energies of the *pure localized single and double reference bonds*. In the field of inorganic chemistry, the question is: where is it possible, with reasonable safety, to find such standards? The tremendous discussions found in the literature over the years — and still today — regarding the nature of the boron-nitrogen bond, for example, lend point to this question.

To sum up, it appears clearly that the concept of *resonance energy*, generally so useful in organic chemistry, can not readily be extended to the general field of ring systems, the more so because its calculations, in the simplest of inorganic cyclic molecules, is strongly dependent on the chosen degree of accuracy and on the personal selection of standards.

This difficulty has furthermore, been felt, even on the level of *pure organic* chemistry, and this is perhaps why the idea of defining aromaticity by the *ring current* physical concept has been so successful [52].

The introduction of this concept is generally attributed to J. A. Pople who used it to explain the N.M.R. deshielding of the benzene ring proton with respect to the ethylene proton [53]. An aromatic compound would be defined, in such conditions, as a molecule which will sustain an induced ring current, the magnitude of which is a function of the ability of π electrons to be delocalized and hence a quantitative measure of *aromaticity*.

It must be noted — and this does not in the least diminish Pople's merits — that Pauling had, about twenty years before, implicitly invoked the same concept to explain the unusually high magnetic susceptibility of the simplest of the arenes [54], and we shall discuss further the general use we may make of the Pauling-Pople ring current concept for all the magnetic properties of matter.

Despite the numerous studies brought to a successful issue by utilizing the ring current notion [55,56,57], it is not free from limitations [58,59,60]. It was, indeed, discovered that the basic deshielding we have explained above does not depend only on the *molecular* ring current but also on *induced* atomic currents [61,62]. Moreover, it was demonstrated that it is quite impossible to link such a concept to either the resonance energy or the reactivity of a molecule [59,63]. But the most precise attack was probably mounted by J. Musher [64] who argued that the N.M.R. behaviour of "aromatic" compounds might be described as the sum of contributions from *localized* electrons of both σ and π character [60,64]; besides, he emphasizes the fact that the "so-called" ring current is a phenomenon which depends essentially on the *group symmetry* of the molecules and that this restriction greatly weakens the generality of this concept [60,65].

The foregoing is not intended to induce in the minds of readers the idea that the ring current concept is dead. It still remains the most powerful tool for solving the problems we meet when we study the magnetic properties of ring systems and, even if it is possible to explain the same phenomena by another model, it would be unreasonable to discard such a well-defined and appropriate concept. However, it must be said that once this concept is used to describe the relative aromaticity inside a class of molecules, it is not legitimate to compare the aromaticity scale obtained in this way to those that have been proposed on the basis of the resonance energy or reactivity factors. Nevertheless, the word "aromaticity" is in fact used for all these different meanings, in the full knowledge that there are absolutely no connections between these different "observables." We shall see further on that this mistake is even more general. Moreover, experimental evidence of the influence of group symmetry, in perfect agreement with the theoretical viewpoint of Musher, will be developed later concerning the Faraday effect.

Let us come back now to the origin of the ring current concept. We have mentioned that this idea was already present in Pauling's famous 1936 work about the specific diamagnetic susceptibility of aromatic compounds. This sort of work was developed after this by many people, particularly by Pacault [66]. The basic observation in this field is the following: the experimental susceptibiliy of benzene, for example, is 9.8 10^{-6} u.e.m C.G.S. more diamagnetic than the theoretical value which may be calculated using the well-known additivity law of localized bond contributions previously put forward by Pascal and co-workers [67,68] in aliphatic chemistry. It was shown, mainly in the recent works of Laity [69] and Dauben [70] that such an elevated diamagnetic susceptibility may be invoked as a criterion for aromaticity which would reflect unequivocally the presence of a cyclic π-electron delocalization.

The importance of this criterion lies in the fact that it is related to a theoretically well-defined quantity, London diamagnetism, and more precisely to diamagnetic anisotropy, which can be measured. It may indeed be demonstrated that a high π delocalization induces into a cyclic molecule a large anisotropy which can be observed by means of a perturbing agent, a magnetic field, for example, as above, or polarized light, as in the D.R.D. technics of Pacault, Bothorel and co-workers [71].

This last remark gives us the opportunity to develop what is perhaps one of the main criticisms against the use of the word aromaticity: if we postulate that aromaticity is an intrinsic, mysterious, non-observable characteristic of the so-called aromatic compounds, it becomes obvious that all we can do is to collect *the various answers provided by this myth to each of the external perturbations,* and nobody can say whether the "aromaticity scales" so obtained will be consistent or not. Moreover, it must be added that each individual way subconsciously create a picture of the myth which is convenient for him. If one accepts this line of argument, it is clear that aromaticity is not a scientific word at all, but fundamentally an esthetic one (an assertion which is supported by the euphonious character of the word aromaticity).

Perhaps the only way to avoid this confusion in the future will be a purely theoretical one: when the theory allows us to predict quantically the experimental signs of aromaticity, it will be possible to approach the unique entity. While awaiting this problematic happy time, it seems better that, at the present stage, each group of scientists using a given experimental approach should try to define and to measure whatever is measurable by their own technique, in the hope that the data thus accumulated will eventually fit into a unified theory.

We have to include in our criticism the use of dipole moments to determine aromaticity. Many attempts were made, mainly in the field of non-alternant systems [72,73,74], but this method was analysed by Dewar [75] who insisted that "the existence of a large dipole moment cannot in itself be taken as evidence that a molecule has a highly delocalized aromatic character".

To sum up what we have said so far, clearly, the best experimental path to a knowledge of what aromaticity is, is the study of the *molecular characteristics in the ground state of the resting compound*, i.e. the compound as free as possible of any external perturbation. Measuring the bond lengths, for example, will give us the opportunity to analyse the famous *bond alternation aromaticity criterion*.

The best methods for this purpose are, of course, X-ray diffraction, electron diffraction and the microwave technique, each being used for a given state of matter. Many results have been published recently, particularly in the field of the first method [76,77,78,79,80,81,82,83]. These fundamental data may be compared with those given by quantum mechanical calculations of bond orders, and we may quote from this viewpoint the aromaticity index proposed by Julg [84] and the delocalization index of Kemula [85] and Trindle [86]. These relations have been tested on a large set of organic molecules, i.e. *homonuclear* molecules. But a quite insuperable difficulty occurs when it becomes necessary to study the bond alternation in *heteronuclear* systems, i.e. essentially in inorganic chemistry. The concept of bond alternation, so well founded from a fundamental point of view, remains at present much too narrow to be used generally as a criterion of aromaticity. Thus, in the field of inorganic chemistry, on the one hand geometrical data are rare and, on the other hand, accurate overall electron calculations must be used, and these are time-consuming and expensive.

However, the importance of bond alternation in the study of the problem of aromaticity has prompted Heilbronner and Binsch to introduce the concepts of *first- and second-order double bond fixation* [87,88,89]. Binsch suggested that a conjugated π-electron system can be called aromatic if it presents neither strong first-order not second-order double bond fixation [90,91]. This criterion has the advantage of being closely related to a physical phenomenon, the tendency of π electrons to cluster in certain bonds. It is capable of pinpointing the lack of π electron delocalization in a particular structural segment. But, once again, this criterion is only applicable to organic chemistry: it is not at all evident that the behaviour of the boron-nitrogen "double" bond of borazole, for example, is identical to that of the C–C bond in benzene.

The critical compilation in the first part of this paper was developed mainly for the purpose of showing *the importance of the troublesome label that organic people have attached to the word aromaticity*. But there is another field in which this label was so much impressed on people's minds that it was out for many years of the question to study the aromaticity of *nonplanar* compounds, because of the "fundamentally necessary assumption", due to Kekule, that "a cyclic molecule cannot be aromatic if it is not planar"

This belief was founded on the fact that, in the classical benzene series, the π delocalization is due to the overlap of $2p_z$ atomic orbitals which quickly becomes zero as soon as the bending of the molecule amounts to some degrees. But, obviously, the situation may be completely different when the π delocalization is due to a ($2p$, $3d$) overlap, for instance, and it may be conceived in such conditions that planarity is no longer a *sine qua non* for the aromaticity of the molecule. This fundamental realisation has allowed the recent study in our laboratory of the ring current intensity of not exactly planar inorganic systems, such as $(RSN)_3$, $(RSN)_4$ and $(R_2PN)_3$.

In the same way, Winstein has introduced the notion of *homo-aromaticity* which is, in fact, more closely synonymous to π *space conjugation* than to delocalization *along* the ring system. It had been observed for many years – by means of some specific techniques which were classically supposed to measure aromaticity – that molecules like 5.6 diphenyl-naphthacene (or rubrene)[92,93], cyclooctatetraene [94] or bisperitetraphenyl-naphthalene [95] have the specific behaviour of "aromatic" compounds although a π delocalization along the chemical bonds of these molecules seems geometrically quite strictly forbidden. A general interpretation of this phenomenon was given, taking into account the space overlap ($2p_\sigma$, $2p_\sigma$) of carbon atomic orbitals which induces in these molecules a three-dimensional conjugation.

To understand such a curious observation is of great interest, but why did the authors attach the prefix *"homo"* to the word aromaticity in this context? This expression was previously used exclusively to describe *ring* delocalization. Such an attitude of mind looks like a survival of the strong organic stamp we have invoked above and serves to strengthen our conviction that the word aromaticity must be eliminated from the *scientific* vocabulary in order to free scientists (and particularly inorganic chemists) from the constraint of having to refer systematically to the benzene molecule and, moreover, to force these same scientists to be more rigorous in their use of the vocabulary of their own field of research.

The time is coming where the "tower of Babel of Aromaticity" must be destroyed.

It ist also unfortunate that, since benzene is looked upon as a standard, the so-called aromaticity of any molecule must be compared with the same quantity estimated for *one of the very rare cyclic systems which belong to the D_{6h} point group*. Moreover, the recent work of Narten [96] has shown that the actual structure of benzene in the liquid state is locally the same as in the solid state, giving a very nice diagram of X-ray diffraction. Benzene is the only benzenic molecule which possesses such properties in the liquid state, and this fact may be directly connected – as Narten says – with the high symmetry (D_{6h}) of this molecule. Is it reasonable, therefore, to use benzene as a standard when it is becoming more and more evident that this compound is *exceptional*? This undoubtedly underlines one of the main errors concerning aromaticity: the specific properties of π delocalized ring systems seem to depend on their symmetry point group, as theoretically asserted by Musher and semi-empirically demonstrated by ourselves [97, 98]. Thus, a comparison of the "aromaticity" of C_{2v} pyridine, for instance, with that of D_{6h} benzene is meaningless. By way of compensation, it is possible to compare the behavior of all the D_{3h} molecules, from 1.3.5 triazine, borazole, boroxine or trimeric phosphonitrilic chloride to mesitylene.

It must be noted that, even if we consider aromaticity as synonymous with chemical reactivity, people implicitly use the concept of point group: for example, the nitration rates for the *ortho-, meta-* and *para-*positions of a given C_{2v} monosubstituted benzene are actually only comparable between themselves and not at all, from a fundamental point of view, with the nitration rates of D_{6h} benzene itself. This tricky problem becomes even more complicated if we want to compare the nitration rates of different substituted benzenes. This fundamental feature will be discussed again under the Faraday effect, but the merit of such a point of view is that it bridges the dangerous gap which exists between the organic chemists and "the rest".

There is indeed misunderstanding between organic chemists — who think they can trust aromaticity — and those who try to approach the same entity with an open mind. An amusing proof of this is the well-known joke: "tris B-methylborazole *may be* aromatic, being an organic molecule, but not borazole itself, because it is inorganic!".

I think thus, concluding this paragraph, that all the approaches of the myth aromaticity would become perhaps more powerful if the concept of group symmetry was present in their developments.

Let me digress a bit from the schedule of this contribution to come back in few words on the fact that benzene seems finally the *worst standard* for the study of aromaticity. What follows is more a psychological remark than a scientific one but I personaly think that it is of some interest. It is at the least curious that many papers appear to-day in the literature under the general imprint *"aromaticity"* which are concerned with "very accurate *ab initio* calculations" of the electronic structure of some "benzene isomers". What is indeed the "aromaticity" — either in terms of ring current or with any other sense — of prismane? It is clear from this particular remark that it becomes too much easy now to insert under the heading "aromaticity" any paper, on the only condition that its title contains the word benzene!

To conclude this rapid survey of the myth of aromaticity, it is clear that the word itself must be extirpated from the scientific literature, despite its esthetic character, not only because it means nothing for a scientific point of view (aromaticity not being an observable property), but chiefly because it too often hides a dangerous imprecision about *what* is being measured and *why*. Moreover, the use of the word restricts the freedom of scientific reasoning.

If we discard this word, we shall simultaneously free all the people who felt obliged to introduce the prefixes and suffixes discussed above in order to be consistent with the tradition. Then it will become possible to come back in every case to the original spirit of the approach and perhaps to progress to a comprehension of techniques which, today seem so classical.

This has been our conviction for many years and is the reason why we have approached our studies of the magneto-optical behaviour of so-called aromatic compounds (*organic or inorganic*) by defining a new concept: *potential strobilism.*

This idea was put forward at the 3rd International Symposium, "aromaticity, pseudo-aromaticity, anti-aromaticity", held in Jerusalem in April 1970. We noted on this occasion with particular pleasure that our attempt agreed with the positions of many other people.

The second part of this paper describes how we have studied the behaviour of what were *previously* called "aromatic molecules" relative to the Faraday effect (or magnetic rotatory power), with particulary emphasis on the experimental and theoretical information which has led us to discard the notion of aromaticity in favour of a new, well-defined and easily measured phenomenon.

First, some generalities about the basis of the Faraday effect:

It is a well-established fact that magnetic rotatory power is what is commonly called an *additive property of matter.* This assertion rests on the fact that there exists [99, 100, 101, 102] a set of additive bond magnetic rotations by means of which the experimental molecular rotation of a compound may be calculated *a priori,* often with surprising precision. The method is applicable to compounds of carbon, boron, sulphur, nitrogen, phosphorous, etc., with the single stipulation that they contain *only normal covalent, diamagnetic and localized bonds.*

The existence of this set has been confirmed on several occasions by utilizing a semi-empirical approach [103, 104] or by adjusting Daudel's "Théorie des Loges" to the Faraday effect [105]. These calculations suggest that any digression from such a law of additivity necessarily implies the presence of certain structural peculiarities in the molecule. This is true, for instance, of molecules containing a delocalized π-electron system: in this case the experimental value (A) of the magnetic rotation is always distinctly higher than the theoretical value (B), calculated by neglecting the delocalization of electrons. The difference $E = (A) - (B)$ — which we have called the "magneto-optical conjugation excess" — appears [106] to be constant for a given unsubstituted conjugated system, but varies appreciably from one system to another.

In the domain of aliphatic compounds, we have shown [107] that E can be linked linearly to the sum, ΣI_r, of the free valence indices of the corresponding system by the equation

$$D_{al} \mid E_{al} = 111 \, (\Sigma I_r - 1.464) \, .$$

This equation has recently appeared also in the framework of a simplified theory of the Faraday effect [108].

The magneto-optical study of a great number of unsubstituted arenes [109] has enabled us to construct an equation for the cyclic conjugated hydrocarbons that is analogous to that for the aliphatic compounds:

$$D_{ar} \mid E_{ar} \simeq 200 \, (\Sigma I_r - 1.464) \, .$$

It may be noted that the straight lines D_{ar} and D_{al} originally have the same abcissa and that the slope of the former is twice as steep as that of the latter. This means that for a given conjugated system, characterized by a value $(\Sigma I_r)_0$, the magneto-optical excess, E_{ar}, which is observed if the molecule is cyclic, is very much higher than E_{al}, the value which would be observed if the molecule were alicyclic.

$\Delta E = (E_{ar} - E_{al})$ expresses, in a way, the passage from open conjugation to cyclic conjugation. The existence of ΔE seems to be due to the fact that a cyclic molecule offers to the delocalized π electrons the possibility of having their mobility increased by the action of the applied magnetic field. We therefore believe that ΔE expresses the existence, in a cyclic molecule, of the Pauling-Pople ring current.

It is not surprising that such currents may be detected by means of the Faraday effect, this effect being a magnetic property of matter to the same extent as, for instance, diamagnetism or N.M.R., in connection with which, as we have seen, the concept of ring current was mentioned for the first time.

We have shown, in the first part of this contribution, how the ring current concept was classically associated with the notion of aromaticity. To be consistent with our earlier discussion, we have introduced the concept of *potential strobilism* to describe the phenomena which are measured by all the magnetic properties of matter in the case of cyclic molecules which are the seat of a π-electron delocalization, that is of a ring current. This new term is not very pleasing phonetically but it has the merit of deriving from a striking image: C. K. Jørgensen suggested this new word on the basis of the Greek phrase which means "children's merry-go-round", a phrase which nicely evokes the ring current.

Strobilism would then *potentially* exist in the molecule, waiting to be rendered observable by a developer (in the "photographic" sense of the word) which, in our case, is the external magnetic field. If strobilism is a potential property of the molecule at rest, it is conceivable that it must be directly dependent upon the electronic and geometrical characteristics of the molecule in the ground state, e.g. electron distribution, symmetry point group, and so on. We will see immediately how it is possible to verify this and in particular to support, from this viewpoint, Musher's criticism of the ring current concept.

To come back to the Faraday effect: It might be expected from the preceding discussion that any molecule characterized by a given ΣI_r value should have the *same conjugation excess*. If one studies a large number of *mono- and polysubstituted* benzenes with the aim of verifying the above, it becomes clear that this is not so. Table 1 shows some of the E values we have obtained [110]. It may be seen that these are sometimes slightly higher and sometimes consid-

Table 1. *E and G values for some substituted benzenes*

N°	$E(\mu r)$	G
6	208	0.006
1	191	0.031
20	187	0.016
8	183	0.066
21	177	0.037
14	176	0.029
3	175	0.032
4	170	0.083
10	166	0.064
12	152	0.101
9	150	0.071
18	148	0.063
23	147	0.030
16	147	0.059
24	137	0.063
17	130	0.109
19	125	0.081
7	123	0.088
2	91	0.112
11	66	0.144
5	63	0.138
15	59	0.101
13	25	0.169
22	12	0.159

erably lower than the characteristics benzene value ($+182\ \mu r$), even though for all these molecules ΣI_r remains constant, at least as long as this quantity is calculated only along the benzene ring.

What therefore can be the electronic or geometrical factors upon which the conjugation excess of a molecule, and therefore its strobilism, depends so strongly?

The answer may be deduced from an examination of the electronic structures we have calculated for abour sixty diversely substituted benzenes, utilizing a combined L.C.A.O.–U.V.C. method, perfected for this purpose in collaboration with Julg [111], and the so-called "bond-by-bond iteration" method [112]. We have noted that if the bond characters remain very close to that observed for benzene (0.667), the gradient

$$G = \frac{1}{6} \sum_{i=1}^{6} |q_i - q_{i+1}| \ \text{(with } q_7 = q_1\text{)}$$

of the $(\sigma + \pi)$ electronic charges, localized on the six carbon atoms of the ring, diverges more or less definitely (Table 1) from the zero value it displays in benzene.

This charge gradient then is responsible for the potential barriers that appear at the peaks of the ring, barriers which have the effect of decreasing the ability of the delocalized π electrons to generate a ring current.

It appears therefore that the strobilism of a molecule, which obviously depends on the density of the delocalized π electrons, is in fact also — and most markedly — a function of the σ electron density of the molecule.

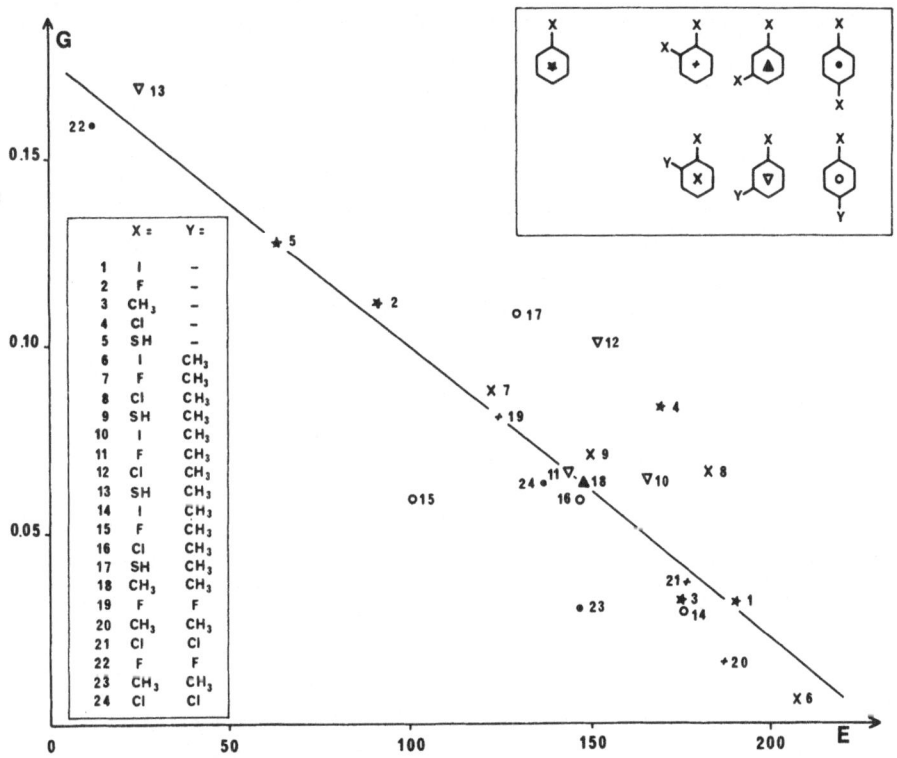

Fig. 1. Linear relation between magneto-optical conjugation excess E and charges gradient G

Thus, we encounter one of the most fundamental objections raised by Musher to the concept of ring current, at least when this concept is developed in its classical form. Moreover, our magneto-optical investigations show that, at the limit, a molecule can possess a very high density of delocalized π electrons without being the seat of a ring current, but only if there is a substantial G gradient. Fig. 1 justifies the reasoning we have just developed: *there exists a very significant linear relation between E and G.*

These results have therefore led us to propose [113] the following definition for the strobilism of a molecule:

A cyclic molecule is strobilic when it contains a distribution of *delocalized π electrons susceptible, in the absence of substantial gradient of localized ($\sigma + \pi$) charges on the atoms of the ring, to be set in ordered movement when submitted to a magnetic field, thus giving rise to a strobilic Pauling-Pople current.*

Strobilism will then either be measured experimentally by means of the Faraday effect (measure of E), or calculated theoretically from the ($\sigma + \pi$) electronic structure (determination of G at equal bond character).

We have therefore two equivalent means of defining the strobilism of a compound relative to others, but on the express condition that they all belong to the same family, that is, on condition that the distribution of the coulomb integrals along the ring *belongs to the same point group.*

This important result is a consequence of the following observation: if one seeks to represent E as a function of G for compounds belonging to the D_{3h} point group (borazoles and boroxines for example [114]) or to the C_{2v} group (furan, pyrrole, thiophene) [97], one also obtains in each case a linear relation. The slope of the representative straight line is similar to that shown in Fig. 1, *but under no circumstances may it be confused with the latter.* There seems to exist a *straight line E = f(G) for each point group, that is, for each symmetry of the distribution of coulomb integrals along the ring.* Here we meet the second objection raised by Musher to the concept of the ring current.

From the preceeding discussion it appears that a *comparison of the strobilism of benzene with that of a molecule not belonging to the D_{6h} group should be avoided.* If we have any right to assign values for the strobilism of various substituted benzenes, studied in relation to benzene itself, it is because the U.V.-consistent Coulomb integrals of carbon atoms bearing an X substituent diverge little, if at all, from the α value when X = R, F, Cl, Br or I. This does not hold for the aminobenzenes, for which the consistency with U.V. compels one to assume the relation $\alpha_{C(NH_2)} = \alpha - 0,6\beta$ [115], a value which, in this peculiar case, confers on the benzene ring a symmetry which is more C_{2v} than D_{6h}.

There is another magneto-optical criterion of strobilism which we proposed in 1967 and which seemed very promising, in view of its simplicity of use. This criterion had been suggested by the N.M.R. experiments carried out by Zimmerman and Foster [116]; these authors showed that the signal of the benzene proton is displaced towards a lower field when benzene is diluted with an inert chemical solvent. Pople thought that this deshileding could be related to the evolution of the ring current during dilution.

An analogous experiment was attempted using the Faraday effect. We noted that the molecular magnetic rotation of benzene decreased — but in a perfectly linear manner — with increasing dilution in a solvent such as n-hexane. Thus we thought that this phenomenon afforded us a simple means of measuring the intensity of the ring current of a molecule and, consequently, its strobilism. We thought it probable — and we believed that we had verified this in a previous study [109] — that the decrease in magnetic rotation would be proportional to the intensity of the ring current.

The first measurements, carried out on fluorobenzene, paradifluorobenzene, chloro- and bromobenzene [109] in solution in n-hexane, allowed us to obtain

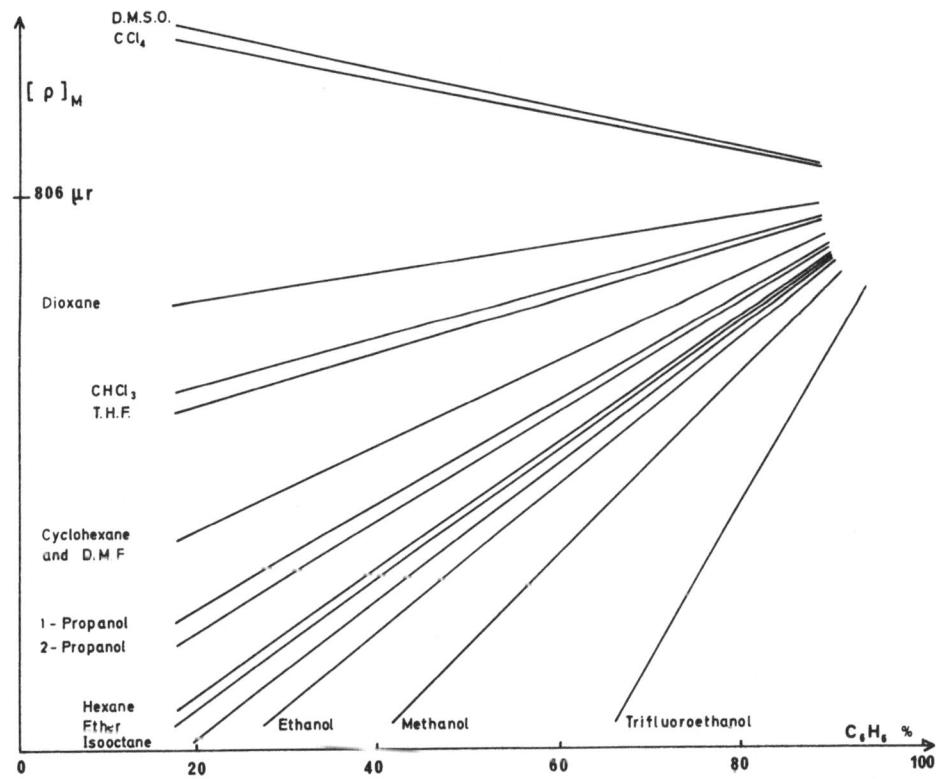

Fig. 2. Variations of the magnetic rotation $(\rho)_M$ of benzene when diluted in some usual solvents

a scale of relative strobilism identical to that deduced from the measurements of E, as well as from calculations of G. But this agreement gradually disappeared as the number of molecules we studied increased. Moreover, we noted that the slope of the straight line obtained for a given strobilic compound — iodobenzene, for instance [117] — varied from one solvent to another in such a pronounced way that the chemical solute-solvent interactions did not suffice to account for it.

It is for this reason that we decided to investigate anew the "criterion for strobilism in solution". We measured the evolution of the rotation of benzene in about fifteen of the usual "inert" solvents. The results are presented in Fig. 2 and Table 2. They call for the following conclusions:

Table 2. *Linear relations between y molecular magnetic rotation of benzene and x concentration in some usual solvents*

Solvent	Equation	r	n_D^{20}	P
Trifluoroethanol	$y = 0.856 x + 730.3$	0.99	1.2910	0.182
Methanol	$y = 0.498 x + 762.8$	0.99	1.3288	0.203
Ethylic Ether	$y = 0.346 x + 773.8$	0.99	1.3526	0.217
Ethanol	$y = 0.399 x + 771.5$	0.99	1.3611	0.221
n-Hexan	$y = 0.338 x + 773.0$	0.99	1.3749	0.229
2-Propanol	$y = 0.294 x + 779.2$	0.98	1.3776	0.230
1-Propanol	$y = 0.287 x + 781.0$	0.99	1.3850	0.234
Isooctane	$y = 0.358 x + 775.0$	0.99	1.3915	0.238
T.H.F.	$y = 0.140 x + 794.0$	0.96	1.4076	0.247
Dioxan	$y = 0.072 x + 801.9$	0.95	1.4224	0.254
Cyclohexan	$y = 0.229 x + 787.2$	0.99	1.4266	0.257
D.M.F.	$y = 0.231 x + 788.9$	0.97	1.4269	0.257
Chloroform	$y = 0.130 x + 798.3$	0.99	1.4433	0.265
CCl$_4$	$y = -0.103 x + 818.5$	0.97	1.4601	0.274
D.M.S.O.	$y = -0.111 x + 819.6$	0.99	1.4780	0.284

1. — The variation of $(\rho)_M$ is linear for a very large series of concentrations, as may be seen from the values of the correlation coefficient r.

2. — The decrease, observed in the case of *n*-hexane, which we had hoped would enable us to establish a scale of strobilism, is not at all general. In certain cases an increase, rather than a decrease, is observed.

3. — As to the sequence in which the investigated solvents could be arranged, it is not clear at first sight whether the macroscopic polarity of the solvents is the determining factor; it is, in fact, surprising that CCl$_4$ and DMSO have almost the same slope.

It is usual to consider solute-solvent interaction phenomena as directly dependent on two main factors: (i) the polarity of the solvent and (ii) the polarizability of the constituents of the interacting system. In view of this, the next logical step seemed to us to investigate whether the variations in $(\rho)_M$ would, in fact, be a function of polarizability rather than of solvent polarity.

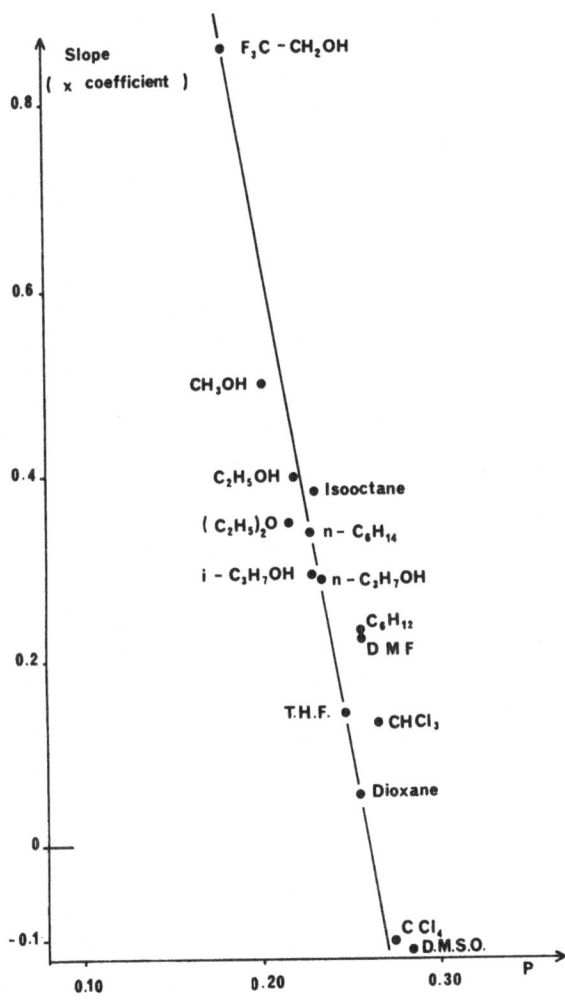

Fig. 3. Linear relation between the slope of the straight lines of Fig. 2 and polarisability P

We calculated for each solvent the molecular polarizability, P, with the Lorentz approximation:

$$P = \frac{(n_D^{20})^2 - 1}{(n_D^{20})^2 + 2}$$

We observed that there exists (Fig. 3) an almost linear relation between the value of P and the slope of the lines in Fig. 2 [118].

It seems clear therefore that the decrease in $(\rho)_M$ is not *directly* dependent on the variation in the ring current intensity of a compound when dissolved in a given solvent. This decrease appears due rather to solute-solvent interactions which modify, to a greater or lesser extent, the *internal electric field* of the medium that is submitted to an *external* magnetic field.

One is compelled to conclude that the study of the magneto-optical behaviour of strobilic molecules in solution does not seem to be directly utilizable as a criterion of strobilism. But the main interest of this conclusion is that it may be generalized to other magnetic properties of matter, and particularly to the N.M.R. field; so many people have tried to measure the "aromaticity" of a molecule by N.M.R. studies in solution.

It may be noted that this concluion has been proposed simultaneously by our team [118] and by that of Falk *et al.* [119].

To sum up this second part of our contribution, we propose to describe the behaviour of the cyclic delocalized molecules in all the magnetic properties of the matter (i.e. N.M.R., diamagnetism, Faraday effect) by the concept of *potential strobilism*. This depends directly upon the electronic and geometrical characteristics of the ground state: electron $(\sigma + \pi)$ distribution, symmetry point group, etc.

Moreover, it is noteworthy that the magnetic properties as a whole give the same strobilic scale for many molecules belonging to the same point group, D_{3h}, for example [120,121], and that this scale is in perfect agreement with the order of strobilism given by quantum chemistry on the basis of the calculation of G gradient values. *I think, in conclusion, that potential strobilism is a new, well-defined concept which can be determined by measuring in a suitable way a given class of physical characteristics, that is, N.M.R., Diamagnetism and the Faraday effect, which are all magnetic properties of matter.*

To conclude this paper, I should like to come back for the last time to the general problem we have discussed and to present a visual image of the situation in which we find ourselves. Chemists and physicists are at present in the middle of a cavern which Plato would not have disavowed: they observe on the walls of the cavern *certain shadows* resulting from the lighting of an unknown subject (aromaticity) by the different sources of light represented by their various chemical or physical techniques of observation: an agreeable odour, an aptitude to nitration and sulphonation, a ring current, a magneto-optical excess, a diamagnetic anisotropy, a resonance energy, a J.V. bathochromic effect, and even a mathematical term. The question is: Do these shadows all belong to *the same* invisible myth (or reality)? and the answer is: "Nobody knows at present".

Therefore, it is indispensable at the present stage that each investigator should try to define and measure whatever is measurable with his technique; we may only hope that the data thus accumulated will eventually fit into a unified concept and a unified theory.

References

1) Kekule, A.: Bull. Soc. Chim. France. *3*, 98 (1865).
2) Peters, D.: J. Chem. Soc. 1274 (1960).
3) Robinson, R.: Tetrahedron *3*, 323 (1958).
4) Lloyd, D. M.: Proceedings. Contribution to the International Symposium on Aromatic, Pseudo-Aromaticity, Anti-Aromaticity, Jerusalem 1970, p. 85.
5) Kekule, A.: Bull. Acad. Roy. Belg. *19*, 551 (1865); Annalen *137*, 129 (1866).
6) Cf. for example: Craig, D. P.: Non-Benzenoid Aromatic Compounds (D. Ginsburg, ed.), p. 1. New York: Interscience Publishers 1959.
7) Thiele, J.: Ber. Deut. Chem. Ges. *33*, 660 (1900).
8) Cf.: Ginsburg, D.: Non-Benzenoid Aromatic Compounds. New York: Interscience Publishers Inc. 1959; Lloyd, D. M.: Carbocyclic Non-Benzenoid Aromatic Compounds. London: Elsevier 1966.
9) Bergmann, E. D., Agranat, I.: Proceedings. Introduction to the International Symposium on Aromaticity, Pseudo-Aromaticity, Anti-Aromaticity, Jerusalem 1970, p. 9.
10) Craig, D. P.: J. Chem. Soc. 3175 (1951).
11) Craig, D. P.: Theoretical Organic Chemistry. Papers presented to the Kekule Symposium, p. 20. London: Butterworths Scientific Publications 1959.
12) Badger, G. M.: Aromatic Character and Aromaticity. New York: Cambridge Univ. Press 1969.
13) Lloyd, D. M., Marshall, D. R.: Chem. Ind. (London) 1760 (1964).
14) Dewar, M. J. S.: Advances in Chemical Physics, Vol VIII, p. 65 (L. Prigogine, ed.), New York: Interscience Publishers 1965.
15) Breslow, R.: Chem. in Britain *4*, 100 (1968); Angew. Chem. Intern. Ed. *7*, 565 (1968).
16) Winstein, S.: Aromaticity, Special Publication n° 21, p. 5. London: Chemical Society 1967.
17) Winstein, S.: Quart. Rev. *23*, 141 (1969).
18) Clark, D. T., Armstrong, D. R.: Chem. Comm. 850 (1969).
19) Hückel, E.: Z. Physik *70*, 204 (1931); Z. Physik *76*, 628 (1932); Physical Society, International Conference on Physic, London, Vol. II, 9 (1935); Z. Elektrochem. *43*, 752 (1937); Grundzüge der Theorie ungesättigter und aromatischer Verbindungen. Berlin: Verlag Chemie 1938.
20) Streitwieser, A., Jr.: Molecular Orbital Theory for Organic Chemists. New York: John Wiley and Sons 1961.
21) Breslow, R.: Chem. Eng. News *43*, 90 (1965).
22) Dewar, M. J. S.: Aromaticity, Special Publication n° 21, p. 177. London: Chemical Society 1967.
23) Coulson, C. A., Longuet-Higgins, H. C.: Proc. Roy. Soc. *A 191*, 32 (1947); Proc. Roy. Soc. *A 192*, 16 (1947); Proc. Roy. Soc. *A 193*, 447, 456 (1948); Proc. Roy. Soc. *A 195*, 188 (1948).

J.-F. Labarre and F. Crasnier

24) Longuet-Higgins, H. C.: J. Chem. Phys. *18*, 265, 275, 283 (1950).
25) Dewar, M. J. S.: J. Am. Chem. Soc. *74*, 3341, 3345, 3350, 3357 (1952).
26) Dewar, M. J. S.: Tetrahedron, Suppl. 8, Part 1, 75 (1966).
27) Dewar, M. J. S.: The Molecular Orbital Theory of Organic Chemistry, p. 191. New York: McGraw-Hill 1969.
28) Osohika, Y.: J. Phys. Soc. Japan *12*, 1238 (1957).
29) Longuet-Higgins, H. C., Salem, L.: Proc. Roy. Soc. *A 251*, 171 (1959).
30) Platt, J. R.: J. Am. Chem. Soc. *76*, 1448 (1954).
31) Baker, W., Mac Omie, J. F. W.: Non-Benzenoid Aromatic Compounds (D. Ginsburg, ed.), p. 477. New York: Interscience Publishers 1959.
32) Vol'pin, M. E.: Russ. Chem. Rev. *29*, 129 (1960).
33) Trost, B. M., Nelsen, S. F., Britelli, D. R.: Tetrahedron Letters 3959 (1967); Trost, B.M.: J. Am. Chem. Soc. *91*, 918 (1969).
34) Pople, J. A.: Trans Faraday Soc. *49*, 1375 (1953).
35) Brickstock, A., Pople, J. A.: Trans. Faraday Soc. *50*, 901 (1954).
36) Fukui, K., Imamura, A., Yonezawa, T., Nagata, C.: Bull. Chem. Soc. Japan *33*, 1591 (1960).
37) Gompper, R., Seybold, G.: Angew. Chem. Intern. Ed., *7*, 824 (1968); Hafner, K. Tappe, T.: Angew. Chem. Intern. Ed. *8*, 593 (1969); Bergmann, E. D.: Chem. Rev. *68*, 41 (1968).
38) Lloyd, D. M., Wasson, F. I.: Chem. Ind. 1559 (1963); J. Chem. Soc. (C), 1086 (1966); Kursanov, D. N., Baranetskaia, N. K., Setkina, V. N.: Dokl. Akad. Nauk SSSR *113*, 116 (1957).
39) Berson, J. A., Evleth. E. M., Hamlet, Z.: J. Am. Chem. Soc. *87*, 2887 (1965); Gompper, R., Weiss, R.: Angew. Chem. Intern. Ed. *7*, 296 (1968).
40) Dewar, M. J. S., Jones, R.: J. Am. Chem. Soc. *90*, 2137 (1968); Davis, F. A., Dewar, M. J. S., Jones, R., Worley, S. D.: J. Am. Chem. Soc. *91*, 2094 (1969).
41) Nakjima, T.: Molecular Orbitals in Chemistry, Physics and Biology (Lowdin, P.O., Pullmann, B., eds.), p. 451. New York: Academic Press 1964.
42) Nakajima, T., Katagiri, S.: Bull. Chem. Soc. Japan *35*, 910 (1962).
43) Nakajima, T., Katagiri, S.: Mol. Phys. *7*, 149 (1964).
44) Julg, A., Francois, P.: Comptes rendus Acad. Sc. (Paris) *258*, 2067 (1964).
45) Chalvet, O., Daudel, R., Kaufman, J. J.: J. Phys. Chem. *68*, 490 (1964).
46) Dewar, M. J. S., Gleicher, G. J.: J. Am. Chem. Soc. *87*, 685 (1965).
47) Dewar, M. J. S., Gleicher, G. J.: J. Am. Chem. Soc. *87*, 692 (1965).
48) Dewar, M. J. S., Llano, C, de: J. Am. Chem. Soc. *91*, 789 (1969).
49) Watanabe, H., Ito, K., Kubo, M.: J. Am. Chem. Soc. *82*, 3294 (1961); Hoffmann, R.: J. Chem. Phys. *40*, 2474 (1964); Bochvar, D. A., Bagatur'yants, A. A.: Izvest. Akad. Nauk SSSR., Otdel. Khim. Nauk. 785 (1963).
50) Brown, D. A., MacCormack, C. G.: J. Chem. Soc. 5385 (1964).
51) Davies, D. W.: Trans Faraday Soc. *56*, 1713 (1960).
52) Elvidge, J. A., Jackman, L. M.: J. Chem. Soc. 859 (1961); Elvidge, J. A.: Chem. Comm. 43 (1965); Hall, G. G., Hardisson, A., Jackman, L. M.: Tetrahedron *19*, Suppl. 2, 101 (1963).
53) Pople, J. A.: J. Chem. Phys. *24*, 1111 (1956).
54) Pauling, L.: J. Chem. Phys. *4*, 673 (1936).
55) Sondheimer, F., Calder, I. C., Elix, J. A., Gaoni, Y., Garratt, P. J., Grohmann, K., Di Maio, G., Mayer, J., Sargent, M. V., Wolovsky, R.: Aromaticity, Special Publication n° 21, p. 75. London: Chemical Society 1967.
56) Boekelheide, V., Phillips, J. B.: J. Am. Chem. Soc. *89*, 1695 (1967).
57) Phillips, J. B., Molyneux, R. J., Sturm, E., Boekelheide, V.: J. Am. Chem. Soc. *89*, 1704 (1967).

52

58) Musher, J. I.: Advances in Magnetic Resonance (J. S. Waugh, ed.), vol II, p. 177. New York: Academic Press 1966.
59) Abraham, R. J., Thomas, W. A.: J. Chem. Soc. (B) 127 (1966).
60) Musher, J. I.: J. Chem. Phys. 43, 4081 (1965); Batuev, M.: Zh. Obshch. Khim. 28, 3147 (1958).
61) Pople, J. A.: J. Chem. Phys. 41, 2559 (1964).
62) Ferguson, A. F., Pople, J. A.: J. Chem. Phys. 42, 1560 (1965).
63) Gaidis, J. M., West R.: J. Chem. Phys. 46, 1218 (1967).
64) Musher, J. I.: J. Chem. Phys. 46, 1219 (1967).
65) Cheshko, F. F.: Zh. Obshch. Khim. 31, 687 (1961).
66) Pacault, A.: Ann. Chim. Fr. 527 (1946).
67) Pacault, A., Lumbroso, N., Hoarau, J.: Conf. de Broglie, Ed. Revue d'Optique théor. et Instr., Paris (1953); ibid., Cahiers de Physique (Paris) 43, 54 (1953).
68) Pascal, P., Gallais, F., Labarre, J.-F.: Comptes rendus Acad. Sc. Paris 256, 335 (1963).
69) Laity, J. L.: Diss. Abstr. Int. B. 30, 559 (1969).
70) Dauben, H. J., jr., Wilson ,J. D., Laity, J. L.: J. Am. Chem. Soc. 90, 811 (1968); J. Am. Chem. Soc. 91, 1991 (1969).
71) Pacault, A.: Proceedings. Contribution to the 3rd International Symposium on Aromaticity, Pseudo-Aromaticity, Anti-Aromaticity, Jerusalem 1970, p. 39.
72) Bergmann, E. D., Agranat, I.: Chem. Comm. 512 (1965).
73) Nakajima, T., Kohda, S., Tagiri, A., Karazawa, S.: Tetrahedron 23, 2189 (1967).
74) Minkin, V. I., Osipov, O. A., Zhdanov, Y. A.: Dipole Moments in organic chemistry (W. E. Vaughan, ed.). New York: Plenum Press 1970.
75) Dewar, M. J. S.: The Molecular Orbital Theory of Organic Chemistry, p. 181–182. New York: McGraw-Hill 1969.
76) Veno, M., Murata, Ik, Kitahara, Y.: Tetrahedron Letters 2967 (1965).
77) Bergmann, E. D., Agranat, I.: Tetrahedron Letters 1275 (1966).
78) Kennard, O., Watson, D. G., Fawcett, J. K., Ann Kerr, K., Romer, C.: Tetrahedron Letters 3885 (1967).
79) Shimanouchi, H., Ashida, T., Sasada, Y., Kakudo, M., Murata, I., Kitahara, Y.: Tetrahedron Letters 61 (1967).
80) Shimanouchi, H., Sasada, Y., Ashida, T., Kakudo, M., Murata, I., Kitahara, Y.: Acta Cryst. 25 B, 1890 (1969).
81) Ammon, H. L.: Tetrahedron Letters 3305 (1969).
82) Dobler, M., Dunitz, J. D.: Helv. Chim. Acta 48, 1429 (1965).
83) Bregman, J., Hirshfeld, F. L., Rabinovitch, D., Schmidt, G. M.: Acta Cryst. 19, 227 (1965).
84) Julg, A., Francois, P.: Theoret. Chim Acta (Berlin) 7, 249 (1967).
85) Kemula, W.: Tetrahedron Letters 5135 (1968).
86) Trindle, C.: J. Am. Chem. Soc. 91, 219 (1969).
87) Binsch, G., Heilbronner, E., Murrell, J. N.: Mol. Phys. 11, 305 (1966).
88) Binsch, G., Heilbronner, E.: Structural Chemistry and Molecular Biology (A. Rich and N. Davidson, eds.), p. 815. San Francisco: W. H. Freeman and Co. 1968.
89) Binsch, G., Heilbronner, E.: Tetrahedron 1215 (1968).
90) Binsch, G., Tamir, I., Hill, R. D.: J. Am. Chem. Soc. 91, 2446 (1969); Binsch, G. Tamir, I.: J. Am. Chem. Soc. 91, 2450 (1969).
91) Binsch, G.: Proceedings. Contribution to the 3rd International Symposium on Aromaticity, Pseudo-Aromaticity, Anti-Aromaticity, Jerusalem 1970, p. 25.
92) Douris, R. G.: Ann. Chim. Fr. 4, 479 (1959).
93) Dufraisse, C., Lepage, Y.: Compt. Rend. Acad. Sc. (Paris) 258, 1507, 5447 (1964).
94) Labarre, J.-F., Chalvet, O.: Tetrahedron Letters 5053 (1967).

95) Chalvet, O., Daudel, R., Evrard, G., Grivet, J-P., Heilbronner, E., Kottis, P., Lava-
lette, D., Muel, B., Straub, P. A., Meersche, M. van: J. Mol. Structure 5, 111 (1970).
96) Narten, A. H.: J. Chem. Phys. 48, 1630 (1968).
97) Devanneaux, J., Labarre, J-F.: J. Chim. Phys. 66, 1780 (1969).
98) Laberre, J-F., Gallais, F.: Proceedings. Contribution to the 3rd International Symposium
on Aromaticity, Pseudo-Aromaticity, Anti-Aromaticity, Jerusalem 1970, p. 48.
99) Gallais, F., Voigt, D.: Bull. Soc. Chim. Fr. 70 (1960).
100) Gallais, F., Voigt, D., Labarre, J-F.: Bull. Soc. Chim. Fr. 2157 (1960).
101) Gallais, F.: Rev. Chim. Min. Fr. 6, 71 (1969).
102) Labarre, J-F., Gallais, F.: Usp. Khim XL, 654 (1971).
103) Daudel, R., Gallais, F.: Rev. Chim. Min. Fr. 6, 61 (1969).
104) Gallais, F., Labarre, J-F., Voigt, D., Loth, Ph de: J. Chim. Phys. 63, 1175 (1966); Labarre, J-F., Labarre, M-C.: J. Chim. Phys. 64, 1670 (1967).
105) Daudel, R., Smet, P., Gallais, F.: Int. J. Quantum Chem. 1, 873 (1967).
106) Labarre, J-F., Gallais, F.: Compt. Rend. Acad. Sc. (Paris) 253, 1935 (1961); Labarre,
J-F.: Ann. Chim. Fr. 8, 45 (1963).
107) Gallais, F., Labarre, J-F.: J. Chim. Phys. 61, 717 (1964).
108) Labarre, J-F.: J. Chim. Phys. 66, 1155 (1969).
109) Labarre, J-F., Loth, Ph. de, Graffeuil, M.: J. Chim. Phys. 63, 460 (1966).
110) Labarre, J-F., Crasnier, F., Faucher, J-P.: J. Chim. Phys. 63, 1088 (1966).
111) Labarre, J-F., Julg, A., Crasnier, F.: Compt. Rend. Acad. Sc. (Paris) 261, 4419 (1965)
112) Gallais, F., Voigt, D., Labarre, J-F.: J. Chim. Phys. 62, 761 (1965).
113) Labarre, J-F., Crasnier, F.: J. Chim. Phys. 64, 1664 (1967).
114) Labarre, J-F., Graffeuil, M., Gallais, F.: J. Chim. Phys. 65, 638 (1968); Labarre, J-F.,
Graffeuil, M., Faucher, J-P., Pasdeloup, M., Laurent, J-P.: Theor. Chim. Acta 11, 423
(1968).
115) Graziana, B.: Ph. D. Thesis, Toulouse (1969).
116) Zimmermann, J. R., Foster, M. R.: J. Phys. Chem. 61, 282 (1957).
117) Bonnafous, M., Graziana, B., Crasnier, F., Labarre, J-F.: J. Chim. Phys. 66, 462 (1969).
118) Labarre, J-F., Moezi, R., Keruzore, J-F.: J. Chim. Phys. 66, 2010 (1969).
119) Bauer, K., Eberhardt, H., Falk, H., Haller, G., Lehner, H.: Monatsch. Chem. 101, 469
(1970).
120) Labarre, J-F.: J. Chim. Phys. 67, 2463 (1970).
121) Pasdeloup, M.: Ph. D. Thesis, Toulouse (1971); Cros, G.: Ph. D. Thesis, Toulouse (1971).

Received February 16, 1971

SPRINGER-VERLAG
BERLIN · HEIDELBERG · NEW YORK

Organometallic Compounds

Methods of Synthesis, Physical Constants and Chemical Reactions
Second Edition. Covering the Literature from 1937 to 1964
Edited by **Michael Dub**

Volume I: ## Compounds of Transition Metals

Edited by **Michael Dub,** Central Research Department,
Monsanto Company
XVIII, 828 pages. 1966. Cloth DM 108,—

Volume II: ## Compounds of Germanium, Tin and Lead

Including Biological Activity and Commercial
Application
Edited by **Richard W. Weiss,** Organic Division,
Monsanto Company
XX, 697 pages. 1967. Cloth DM 108,—

Volume III: ## Compounds of Arsenic, Antimony and Bismuth

Edited by **Michael Dub,** Central Research Department,
Monsanto Company
XX, 925 pages. 1968. Cloth DM 108,—

Formula Index to Volumes I to III

Prepared by Michael Dub and Richard W. Weiss, Monsanto Company

Second edition

VII, 343 pages. 1969
Cloth DM 72,—

This formula index contains the compounds of all three volumes. The molecular formulae show metal atoms first, followed by carbon, hydrogen, and other nonmetal atoms arranged alphabetically. The mono-metallic and homopolymetallic compounds are followed by hetero-bimetallic, -trimetallic and -poly-metallic compounds. Heterometallic compounds are listed under each metal.